Annals of Mathematics Studies
Number 202

Arithmetic and Geometry

Ten Years in Alpbach

EDITED BY

Gisbert Wüstholz and Clemens Fuchs

PRINCETON UNIVERSITY PRESS

PRINCETON AND OXFORD

2019

Published by Princeton University Press
41 William Street, Princeton, New Jersey 08540
6 Oxford Street, Woodstock, Oxfordshire OX20 1TR

press.princeton.edu

Library of Congress Control Number:2019943220
ISBN 978-0-691-19378-6
ISBN (pbk.) 978-0-691-19377-9

British Library Cataloging-in-Publication Data is available

Editorial: Vickie Kearn, Susannah Shoemaker, and Lauren Bucca
Production Editorial: Nathan Carr
Text Design: Leslie Flis
Jacket/Cover Design: Leslie Flis
Production: Jacquie Poirier
Publicity: Matthew Taylor and Katie Lewis

This book has been composed in LaTeX

Printed on acid-free paper. ∞

Printed in the United States of America

10 9 8 7 6 5 4 3 2 1

Contents

Preface

GISBERT WÜSTHOLZ

IN THE LAST DECADES one of the most active and flourishing parts in mathematics was arithmetic and Diophantine geometry, partly encouraged by Faltings's celebrated proof of the Mordell conjecture together with foundational work of Arakelov, Gillet, and Soulé on arithmetic intersection theory. Also, the conjectures of André and Oort had much influence, and were later extended by Pink and others on special points on Shimura and mixed Shimura varieties. We could see further important progress in the theory of motives in connection with algebraic independence properties of multi-zeta values as well as polylogarithms obtained by Deligne, Goncharov, F. Brown, and others. Another field of research was opened by some important work of Kontsevich and Zagier who studied the period ring of rational integrals. A striking consequence was a very far-reaching conjecture about the structure of this ring in terms of generators and relations. Periods became central research topics, in particular the study of periods in the p-adic domain which has made very substantial progress in the last ten years as well as also p-adic modular forms. These topics can be seen as a part of p-adic Hodge theory initiated by Faltings in the '80s and developed very rapidly in recent years. It became clear that classical Hodge theory and p-adic Hodge theory are of very different but *pari passu* nature.

To find the solutions of Diophantine equations, effectively or only qualitatively, has been a central subject in Diophantine geometry over centuries. We mention in this context only the work of Baker on logarithmic forms which found applications for various classes of Diophantine equations and problems.

More and more deep methods from arithmetic algebraic geometry were introduced since Baker's work to attack what were previously seemingly inaccessible problems. Highlights were Faltings's proof of the Mordell conjecture and Wiles's proof of Fermat's last theorem, the first giving only a qualitative result, the latter giving the most pure quantitative result. This success motivated researchers to consider again old and almost forgotten methods to find effective solutions for certain types of Diophantine equations. A prototype is the so-called Chabauty's method which uses p-adic methods for finding solutions effectively in certain cases. One initial step and important ingredient was Coleman's p-adic integration for further deep studies to find solutions.

In 2007 we started with a series of yearly summer schools in Alpbach in the Austrian mountains organized by the research groups in arithmetics and geometry at ETH Zurich (ETHZ) and the University of Zurich (UZh). The aim was to study such topics as mentioned and to make them accessible to graduate students and young researchers. The concept was to choose a subject and to organize lectures by young people to learn up-to-date new theories. At the beginning the participants came mostly from Zurich, Berlin, Mainz, and Freiburg. Over the years and without advertisement, more and more international participants joined the originally more local group. The very high standard of the program and the speakers made Alpbach internationally recognized.

This year we modified our concept and asked international top researchers to give mini-courses on a subject of their research. The organization and selection of the subject and of the speakers was by J. Ayoub and A. Kresch (UZh), G. Wüstholz (ETHZ), and C. Fuchs, formerly ETHZ and now University of Salzburg. We formed several groups of young participants to produce notes of the courses and this created a very stimulating working atmosphere.

The Alpbach Summerschool 2016 was the celebration of the tenth session with outstanding speakers covering very different research areas in arithmetic and Diophantine geometry. The topics were "Local Shimura Varieties" by Peter Scholze, "Hyperelliptic Continued Fractions and Generalized Jacobians" by Umberto Zannier, and "Faltings Heights and L-functions" by Shou-Wu Zhang. This monograph consists of notes written and substantially complemented by groups of young researchers, and the aim of this book is to make the minicourses available to a larger community. We believe that the quality of the courses and of the notes is exceptional and justifies their publication. We should point out that the notes were written by different collectives and were authorized by the lecturers.

The ten summer schools would not have been possible without the enthusiasm of the organizers for up-to-date arithmetic and Diophantine progress. Over the years young researchers made great efforts to get into difficult theories and results, in particular the three 2016 teams which made these lecture notes possible. They did a great job and we hope that the reader can participate in our love for beautiful mathematics.

The team of the Romantik Hotel Böglerhof provided us with excellent service which was appreciated by all the participants over the years so that each year at the end of our stay in Alpbach we were looking forward to being there the year after.

We were supported regularly by the Zurich Graduate School in Mathematics, the Forschungsinstitut für Mathematik at ETHZ, and in recent years also by the FWF, the Austrian science foundation (grant no. P24574).

Gisbert Wüstholz
Alpbach 2017

Arithmetic and Geometry

Chapter One

Introduction

THE FIRST PART is notes of a course given by Peter Scholze on local Shimura varieties and contains very recent results of Scholze's. They are part of the Langlands program and deal with the local Langlands conjecture.

Local Langlands conjecture predicts a correspondence between representations of the Weil group of the field of p-adic numbers $\mathbb{Q}_p$ and (possibly, infinite-dimensional) representations of p-adic reductive groups. The Weil group $W_{\mathbb{Q}_p}$ of $\mathbb{Q}_p$ consists of all elements in the absolute Galois group of $\mathbb{Q}_p$ whose restriction to the maximal unramified extension of $\mathbb{Q}_p$ equals an integral power of Frobenius.

To be more precise, fix a reductive algebraic group G over $\mathbb{Q}_p$ and a prime number ℓ different from p. On one side of the local Langlands correspondence, one considers irreducible smooth representations π over the field $\bar{\mathbb{Q}}_\ell$ of the p-adic group $G(\mathbb{Q}_p)$. On the other side of the correspondence, one considers L-parameters of G, which are continuous homomorphisms from $W_{\mathbb{Q}_p}$ to the Langlands dual group ${}^L G$ over $\bar{\mathbb{Q}}_\ell$ that satisfy certain properties. In particular, if G is split, then an L-parameter is the same as a continuous homomorphism from $W_{\mathbb{Q}_p}$ to the connected component ${}^L G^\circ$ of the Langlands dual group (up to conjugation). Local Langlands correspondence associates conjecturally to π its L-parameter $LLC(\pi)$. This association should satisfy certain special properties.

First, consider the group $G = \boldsymbol{GL}_n$. In this case, the local Langlands conjecture was proven by Harris-Taylor and Henniart. The group $\boldsymbol{GL}_n$ is split and ${}^L G^\circ = \boldsymbol{GL}_n$. Thus in this case, an L-parameter is the same as an n-dimensional representation of $W_{\mathbb{Q}_p}$ over $\bar{\mathbb{Q}}_\ell$. A linking bridge between representations of $\boldsymbol{GL}_n(\mathbb{Q}_p)$ and of $W_{\mathbb{Q}_p}$ is provided by the following general construction: let H be an (infinite-dimensional) $\bar{\mathbb{Q}}_\ell$-vector space together with commuting actions of $W_{\mathbb{Q}_p}$ and $\boldsymbol{GL}_n(\mathbb{Q}_p)$. Then for any π as above, one obtains a $W_{\mathbb{Q}_p}$-representation $\mathrm{Hom}_{\boldsymbol{GL}_n(\mathbb{Q}_p)}(\pi, H)$. (Actually, this will not yet be the n-dimensional representation $LLC(\pi)$ of $W_{\mathbb{Q}_p}$, but it allows one to reconstruct $LLC(\pi)$ with the help of the (non-conjectural) Jacquet-Langlands correspondence.)

Of course, one needs to make a very particular choice of H so that it gives the local Langlands correspondence. According to Harris and Taylor, the needed H is given by ℓ-adic cohomology of the Lubin-Tate space. This is a (pro-)rigid analytic space over the p-adic completion L of the unramified extension of $\mathbb{Q}_p$. The Lubin-Tate space has the following modular sense: it parametrizes deformations of the one-dimensional formal group law of height n over $\bar{\mathbb{F}}_p$ considered up to quasi-isogenies with a p-adic level structure, that is, with a trivialization of the p-adic Tate module of a deformation.

Now let G be a reductive group over $\mathbb{Q}_p$ and, for simplicity, assume that G is split. Then the L-parameter $LLC(\pi)$ is given by a homomorphism $W_{\mathbb{Q}_p} \to {}^L G^\circ$. Thus it is natural to detect $LLC(\pi)$ by describing its compositions $r \circ LLC(\pi)$ with various (irreducible, finite-dimensional, and algebraic) representations r of ${}^L G^\circ$ over $\bar{\mathbb{Q}}_\ell$. Similarly as above, one constructs the representations $r \circ LLC(\pi)$ of $W_{\mathbb{Q}_p}$ with the help of cohomology of certain p-adic spaces endowed with the actions of both $W_{\mathbb{Q}_p}$ and $G(\mathbb{Q}_p)$.

Already in the case $G = \boldsymbol{GL}_n$, one can search for representations of $\boldsymbol{GL}_n$ other than the tautological one. The case of wedge powers of the tautological representation is treated with the help of Rapoport-Zink spaces, which generalize Lubin-Tate spaces. More precisely, for the d-th wedge power of the tautological representation of $\boldsymbol{GL}_n$, one takes ℓ-adic cohomology of the Rapoport-Zink space that has the following modular sense: it parametrizes deformations of the isoclinic p-divisible group of height n and dimension d over $\bar{\mathbb{F}}_p$ considered up to quasi-isogenies with a p-adic level structure on a deformation.

Generalizing this to an arbitrary reductive group G, one replaces a p-divisible group over $\bar{\mathbb{F}}_p$ by purely group-theoretic datum called a local Shimura datum, which consists of a conjugacy class $\bar{\mu}$ of a cocharacter $\mu \colon \mathbb{G}_m \to G$ (possibly, defined over an extension of $\mathbb{Q}_p$) and a Frobenius-twisted conjugacy class b in $G(L)$. The cocharacter μ and the conjugacy class b are assumed to be compatible in a way. Also, one requires μ to be minuscule.

It was predicted by Kottwitz that for each local Shimura datum $(G, b, \bar{\mu})$, there exists a so-called local Shimura variety, which is a (pro-)rigid analytic space. This is now an unpublished theorem, by the work of Fargues, Kedlaya-Liu, and Caraiani-Scholze. Furthermore, ℓ-adic cohomology of the local Shimura variety conjecturally provides a way to reconstruct the representation $r_\mu \circ LLC(\pi)$, where r_μ is the highest weight representation of ${}^L G^\circ$ associated naturally with the cocharacter μ of G.

If the cocharacter μ is not minuscule, then one does not expect the existence of such local Shimura variety in the category of (pro-)rigid analytic spaces. Thus the above approach deals only with minuscule representations, which is a small part of all representations r_μ of ${}^L G^\circ$. Hence it does not allow one to reconstruct the conjectural L-parameter $LLC(\pi)$. To solve this problem, one needs to introduce new geometric objects instead of rigid analytic spaces.

The new geometric objects live in the world of perfectoid spaces. Namely, they are diamonds, that is, algebraic spaces with respect to the pro-étale topology on perfectoid spaces (in particular, rigid analytic spaces are diamonds). Even more, perfectoid spaces allow one to interpret local Langlands correspondence as a geometric Langlands correspondence on a perfectoid curve called a Fargues-Fontaine curve. This opens a new powerful approach to constructing the local Langlands correspondence. Though this program is not yet entirely implemented, substantial progress has been achieved in recent works by Fontaine, Fargues, Scholze, Weinstein, Caraiani, and V. Lafforgue.

The second part, a course by Umberto Zannier, deals with a rather classical theme but from a modern point of view. His course is on hyperelliptic continued fractions and generalized Jacobians. The starting point is the classical Pell equation which he considers over the ring of polynomials over $\mathbb{C}$.

The classical Pell equation is the Diophantine equation $x^2 - dy^2 = 1$, to be solved in integers x and y, where d is some fixed integer. The problem reduces to the case of positive non-square integers d, in which case there are always infinitely many solutions, each of which can be generated from a minimal one. As is well-known today, Pell's equation is strongly related to the theory of continued fraction expansions of real quadratic numbers, as the solutions appear as numerators and denominators of convergents of $\sqrt{d}$. It was Lagrange who proved that indeed there always exists a solution, showing that the sequence of partial quotients of a real number a is eventually periodic if and only if a is algebraic of degree two.

Now let $D \in \mathbb{C}[t]$ be a complex polynomial of even degree and consider the "polynomial Pell equation," $x(t)^2 - D(t)y(t)^2 = 1$. We call D *Pellian* if there are nonzero polynomials $x, y \in \mathbb{C}[t]$ which solve the equation. Similarly, as in the classical setting, one may define the continued fraction expansion of $\sqrt{D(t)}$, viewed as a Laurent series at infinity, proceeding analogously to the classical algorithm, but now replacing the integer part by its polynomial part.

One may ask whether there is again a correspondence between the solvability of the polynomial Pell equation and the continued fraction expansion of $\sqrt{D(t)}$. Indeed, it was already known by Abel and Chebyshev that D is Pellian if and only if the sequence of partial quotients in the continued fraction of $\sqrt{D(t)}$ (which are polynomials now) is eventually periodic. On the other hand, this turns out to be a quite rare phenomenon and, in contrary to the case of real numbers, periodic continued fraction expansions are not limited to square roots of polynomials.

However, Zannier discovered that some periodicity still remains in full generality: The sequence of the degrees of the partial quotients of $\sqrt{D(t)}$ is eventually periodic. In the case that D is squarefree, the proof relies on certain divisor relations in the associated Jacobian variety of the underlying projective curve corresponding to $u^2 = D(t)$ and the application of a variant of the theorem of Skolem-Mahler-Lech for algebraic groups. The case of nonsquarefree D, on the other hand, involves the study of so-called generalized Jacobians.

Among the periodicity of the partial quotients of $\sqrt{D(t)}$, the Pellianity of certain families of polynomials $D_\lambda \in \mathbb{C}(\lambda)[t]$ is investigated as well, such as $D_\lambda(t) = t^4 + \lambda t^2 + t + 1$. One may ask for which specializations of the parameter $\lambda \in \mathbb{C}$ the equation in question has a nontrivial solution. Again, the study relies on a criterion which links the solvability to certain points on the associated (generalized) Jacobians. Anyhow, there are also various further phenomena to observe in the context of continued fraction expansions of Laurent series as well as connections to other related problems, such as *unlikely intersections* in families of Jacobians of hyperelliptic curves or *Padé approximations*.

The notes are organized as follows. After the topic is motivated and some historical background is given in the first section, Sections 3.2–3.4 are dedicated to continued fractions. We recall the method and some related results for the standard setup in Section 3.2. The third section generalizes continued fractions to more general settings, giving some illustrative examples, whereas Subsection 4 explains the procedure for Laurent series in detail. Section 3.5 deals with the continued fraction expansion of $\sqrt{D}$ and gives related results, including the main theorem. In Section 3.6 a criterion for D to be Pellian in terms of a special point of the Jacobian of the underlying curve in the case of squarefree D is given, thereupon Section 3.7 treats the case of D not being squarefree. Various pencils of polynomials, for squarefree and nonsquarefree D_λ, are analyzed as to their Pellianity in these two subsections. Sections 3.8 and 3.9 are dedicated to the mentioned version of the Skolem-Mahler-Lech theorem and the proof of the periodicity of the partial quotients of $\sqrt{D(t)}$. In the appendix, the reader shall find solutions to the exercises given during the lecture.

The theme of the third course by Shou-Wu Zhang originates in the famous Chowla-Selberg formula which was taken up by Gross and Zagier in 1984 to relate values of the L-function for elliptic curves with the height of Heegner points on the curves. Only in recent years has a very significant step been taken by P. Colmez relating L-values for abelian varieties with complex multiplication to the Faltings height of the abelian variety. Building on this work, X. Yuan, Shou-Wu Zhang, and Wei Zhang succeeded in proving the Gross-Zagier formula on Shimura curves and shortly later they verified the Colmez conjecture on average. In the course Zhang presents new interesting aspects of the formula.

Let K be a number field and let A be an abelian variety over K of dimension $g \geq 1$. In the first part, we will define the stable Faltings height $h(A)$ of A. It is a real number associated to A which is invariant under base change. Faltings introduced this invariant in his paper on the Mordell conjecture, Shafarevich conjecture, and Tate conjecture in order to study isogenies of abelian varieties over number fields. The stable Faltings height has since gained momentum as a tool to answer other questions in arithmetic geometry, and has a deep connection with the abc-conjecture, which is a consequence of the following conjecture.

Conjecture 1.1 (Generalized Szpiro Conjecture). *Let K be a number field and let $g \geq 1$ be an integer. Then any abelian variety A over K of dimension g satisfies*

$$h(A) \leq \tfrac{\alpha}{[K:\mathbb{Q}]} \left(\log N_A + \log \Delta_K \right) + \beta, \qquad (1.1)$$

where $\alpha, \beta \in \mathbb{R}$ are constants depending only on $[K:\mathbb{Q}]$ and g. Here Δ_K denotes the absolute discriminant of $K/\mathbb{Q}$ and N_A denotes the norm of the conductor ideal of A/K.

A strong form of this conjecture says that for each real number $\epsilon > 0$ one can take here $\alpha = \frac{g}{2} + \epsilon$ and β depending only on g and ϵ. The generalized Szpiro conjecture has many striking consequences. In particular, it implies an effective version of the Mordell conjecture (assuming α and β are effectively computable), and "no Landau-Siegel zero" which follows from the strong form of the conjecture for CM elliptic curves. We shall discuss several applications of the generalized Szpiro conjecture and we shall also consider a function field analogue: The so-called "Arakelov inequality" which was proved by Arakelov, Faltings, and Parshin.

In the second part, we will consider the case of CM abelian varieties. Let E be a CM field of degree $[E : \mathbb{Q}] = 2g$ and assume that A has CM type $(\mathcal{O}_E, \Phi)$ where Φ is a CM type of E and $\mathcal{O}_E$ is the ring of integers of E. Faltings height computations are especially amenable to these abelian varieties. In particular, Colmez showed that $h(A)$ only depends on the CM type Φ. We may therefore write $h(\Phi)$ to denote this Faltings height. In the case of CM elliptic curves we can in fact say more using the following version of the formula of Lerch-Chowla-Selberg.

Theorem 1.2 (Lerch-Chowla-Selberg). *Suppose that E is an imaginary quadratic field of discriminant $-d < 0$ and let $\eta : (\mathbb{Z}/d\mathbb{Z})^\times \to \{\pm 1\}$ be its quadratic character. Then it holds*

$$h(\Phi) = -\tfrac{1}{2} \frac{L'(\eta, 0)}{L(\eta, 0)} - \tfrac{1}{4} \log d,$$

where $L'(\eta, s)$ is the derivative of the Dirichlet L-function $L(\eta, s)$ of η.

We will discuss several generalizations and reformulations of this formula. In particular we shall consider the following averaged version of a conjecture of Colmez.

Theorem 1.3 (Averaged Colmez Conjecture). *Let F be the maximal totally real subfield of E. Denote by $\Delta_{E/F}$ the norm of the relative discriminant of E/F, and write Δ_F for the absolute discriminant of F. Then it holds*

$$\tfrac{1}{2^g} \sum_\Phi h(\Phi) = -\tfrac{1}{2} \frac{L'(\eta, 0)}{L(\eta, 0)} - \tfrac{1}{4} \log(\Delta_{E/F} \Delta_F)$$

with the sum taken over all distinct CM types Φ of E. Here $L'(\eta, s)$ is the derivative of the L-function $L(\eta, s)$ of the quadratic character η of $\mathbb{A}_F^\times / F^\times$ defined by E/F.

The above theorem was proven by Yuan-Zhang and independently by Andreatta-Goren-Howard-Madapusi Pera. The averaged Colmez conjecture played a key role in proving the André-Oort conjecture for large classes of Shimura varieties, including the moduli spaces A_g of principally polarized abelian varieties of arbitrary positive dimension g. We shall explain the main ideas and concepts of the proof of the averaged Colmez conjecture given by Yuan-Zhang. Their proof involves Shimura curves and it uses the method of Yuan-Zhang-Zhang which they developed to prove generalized Gross-Zagier formulas for Shimura curves.

In the last part, we will then discuss the work of Yun-W. Zhang which can be viewed as a simultaneous generalization for function fields of the Chowla-Selberg formula, the Waldspurger formula, and the Gross-Zagier formula. In particular they studied higher order derivatives of certain L-functions at the center and they proved a formula for unramified cuspidal automorphic representation π of PGL_2 over a function field $F = k(X)$, where X is a curve over a finite field k. In fact they express the r-th central derivative of the L-function (base changed along a quadratic extension E of F) in terms of the self-intersection number of the π-isotypic component of the Heegner-Drinfeld cycle Sht_T^r on the moduli stack Sht_G^r.

Theorem 1.4 (Higher Gross-Zagier formula, Yun-W. Zhang). *There is an explicit positive constant $c(\pi)$ such that*

$$([\mathrm{Sht}_T^r]_\pi, [\mathrm{Sht}_T^r]_\pi) = c(\pi) L^{(r)}(\pi_E, 1/2).$$

Here the moduli stack Sht_G^r is closely related to the moduli stack of Drinfeld shtukas of rank two with r modifications. An important feature of Sht_G^r is that it admits a natural fibration $\mathrm{Sht}_G^r \to X^r$ where X^r is the r-fold self-product of X over k. In the number field case, the analogous spaces (currently) only exist when $r \leq 1$. In the case $r = 0$, the moduli stack Sht_G^0 is the constant groupoid over k given by $\mathrm{Bun}_G(k) \cong G(F)\backslash G(\mathbb{A}_F)/H$ where $\mathbb{A}_F$ is the ring of adèles of F and where H is a maximal compact open subgroup of $G(\mathbb{A}_F)$. The double coset $G(F)\backslash G(\mathbb{A}_F)/H$ has a meaning when F is a number field. In the case when $r = 1$ and $F = \mathbb{Q}$, the counterpart of Sht_G^1 is the moduli stack of elliptic curves which is defined over $\mathbb{Z}$.

Chapter Two

Local Shimura Varieties: Minicourse Given by Peter Scholze

SERGEY GORCHINSKIY AND LARS KÜHNE

THESE ARE NOTES from the minicourse given by Peter Scholze (University of Bonn). The notes were worked out by Sergey Gorchinskiy[1] and Lars Kühne.

2.1 INTRODUCTION

Local Langlands conjecture predicts a correspondence between representations of the Weil group of the field of p-adic numbers $\mathbb{Q}_p$ and (possibly, infinite-dimensional) representations of p-adic reductive groups. Recall that a Weil group is almost the same as the absolute Galois group of $\mathbb{Q}_p$, namely, it consists of all elements in the absolute Galois group whose restriction to the maximal unramified extension of $\mathbb{Q}_p$ acts as an integral power of Frobenius.

One fixes a reductive group G over $\mathbb{Q}_p$ and a prime ℓ different from p. On one side of the correspondence, one considers irreducible smooth representations π over $\bar{\mathbb{Q}}_\ell$ of the p-adic group $G(\mathbb{Q}_p)$. On the other side of the correspondence, one considers L-parameters of G, which are continuous homomorphisms from the Weil group $W_{\mathbb{Q}_p}$ to the Langlands dual group $^L G$ over $\bar{\mathbb{Q}}_\ell$ that satisfy certain properties. In particular, if G is split, then an L-parameter is the same as a continuous homomorphism from $W_{\mathbb{Q}_p}$ to the connected component $^L G^\circ$ of the Langlands dual group. Local Langlands correspondence associates conjecturally to π its L-parameter $LLC(\pi)$. See more detail in Section 2.2.

The goal of these lectures is to describe a program to construct local Langlands correspondence. The construction is based on cohomology of so-called local Shimura varieties and generalizations thereof.

For $G = \boldsymbol{GL}_n$ local Langlands conjecture was proven by Harris-Taylor [25] and Henniart [27].

[1]S. Gorchinskiy served as group leader of the working group "Minicourse Scholze" consisting of Sergey Gorchinskiy, Lars Kühne, Alexandre Puttick, and Alberto Vezzani. The authors are grateful to Alberto Vezzani and Alexandre Puttick for their contribution to the draft version of this text.

The method of Harris and Taylor uses cohomology of a Lubin-Tate space $\mathcal{M}_\infty$. This is a pro-object in the category of rigid analytic spaces over the p-adic field $L = W(\bar{\mathbb{F}}_p)[\frac{1}{p}]$. Lubin-Tate space parametrizes deformations of the one-dimensional formal group law of height n over $\bar{\mathbb{F}}_p$ considered up to quasi-isogenies with a p-adic level structure, that is, with a trivialization of the p-adic Tate module of a deformation.

The group $\boldsymbol{GL}_n(\mathbb{Q}_p)$ acts naturally on $\mathcal{M}_\infty$ and the corresponding action on ℓ-adic étale cohomology $H_{c,\text{ét}}^{n-1}(\mathcal{M}_\infty, \bar{\mathbb{Q}}_\ell)$ commutes with a natural action of the Weil group $W_{\mathbb{Q}_p}$ on this space. Using the representation of $W_{\mathbb{Q}_p} \times \boldsymbol{GL}_n(\mathbb{Q}_p)$ in $H_{c,\text{ét}}^{n-1}(\mathcal{M}_\infty, \bar{\mathbb{Q}}_\ell)$ and also the (non-conjectural) Jacquet-Langlands correspondence, one obtains a construction of the local Langlands correspondence for $\boldsymbol{GL}_n$. More precisely, given a supercuspidal representation π of $\boldsymbol{GL}_n(\mathbb{Q}_p)$, one constructs explicitly the corresponding (irreducible) n-dimensional representation $LLC(\pi)$ of $W_{\mathbb{Q}_p}$.

The approach to local Langlands correspondence via cohomology of Lubin-Tate spaces is explained in Section 2.3. In Section 2.3.1 we start by recalling basics on one-dimensional formal group laws, then we introduce a formal deformation problem and describe properties of the corresponding universal deformation formal scheme. Then we add quasi-isogenies and level structures into our considerations and finally come to a pro-object in the category of rigid analytic spaces, the Lubin-Tate space $\mathcal{M}_\infty$. We discuss the actions of the group $\boldsymbol{GL}_n(\mathbb{Q}_p)$, of the group of units $D^\times$ in the division algebra D over $\mathbb{Q}_p$ of dimension n^2, and of Frobenius on the Lubin-Tate space $\mathcal{M}_\infty$. In Section 2.3.2, we consider ℓ-adic étale cohomology of the Lubin-Tate space and the arising actions of $\boldsymbol{GL}_n(\mathbb{Q}_p)$, $D^\times$, and $W_{\mathbb{Q}_p}$ on this cohomology. This allows us to state the construction of the local Langlands correspondence for $\boldsymbol{GL}_n$.

Let us mention that for $G = \boldsymbol{GL}_n$, one uses implicitly the following particular property: giving a continuous homomorphism from $W_{\mathbb{Q}_p}$ to $^L G^\circ = \boldsymbol{GL}_n$ over $\bar{\mathbb{Q}}_\ell$ is the same as giving an n-dimensional representation of $W_{\mathbb{Q}_p}$ over $\bar{\mathbb{Q}}_\ell$.

Now for an arbitrary reductive group G, it is natural to consider various representations r of $^L G$. Namely, given a supercuspidal representation π of $G(\mathbb{Q}_p)$, one aims to detect the L-parameter $LLC(\pi)$ by constructing the representations $r \circ LLC(\pi)$ of $W_{\mathbb{Q}_p}$ with the help of cohomology of certain p-adic spaces. Already in the case $G = \boldsymbol{GL}_n$ one can ask for representations of $\boldsymbol{GL}_n$ other than the tautological one. The case of wedge powers of the tautological representation is treated with the help of Rapoport-Zink spaces, which generalize Lubin-Tate spaces.

Rapoport-Zink space $\mathcal{M}_{\mathbb{X},\infty}$ parametrizes deformations of a p-divisible group $\mathbb{X}$ of height n and dimension d over $\bar{\mathbb{F}}_p$ considered up to quasi-isogenies with a p-adic level structure on a deformation. Similarly as for Lubin-Tate spaces, we have a representation of $W_{\mathbb{Q}_p} \times \boldsymbol{GL}_n(\mathbb{Q}_p)$ in cohomology $H_{c,\text{ét}}^{n-1}(\mathcal{M}_{\mathbb{X},\infty}, \bar{\mathbb{Q}}_\ell)$. By a conjecture of Kottwitz, now essentially a theorem of Fargues, this allows one to construct $\bigwedge^d LLC(\pi)$ for a supercuspidal representation π of $\boldsymbol{GL}_n(\mathbb{Q}_p)$ provided that $\mathbb{X}$ is the isoclinic p-divisible group of height n and dimension

d (again, using also the Jacquet-Langlands correspondence). The case $d = 1$ is reduced to Lubin-Tate spaces by an equivalence between divisible commutative formal groups and connected p-divisible groups.

It is natural to generalize this to an arbitrary reductive group G. A p-divisible group $\mathbb{X}$ over $\bar{\mathbb{F}}_p$ is replaced now by a local Shimura datum, which consists of a conjugacy class $\bar{\mu}$ of a cocharacter $\mu\colon \mathbb{G}_m \to G$ (possibly, defined over an extension of $\mathbb{Q}_p$) and a Frobenius-twisted conjugacy class b in $G(L)$. The cocharacter μ and the conjugacy class b are assumed to be compatible in a way. More substantially, μ is assumed to be minuscule.

It was predicted by Kottwitz that for each local Shimura datum $(G, b, \bar{\mu})$, there exists a so-called local Shimura variety, which is a pro-object in the category of rigid analytic spaces, with natural properties including the action of the group $G(\mathbb{Q}_p)$. Thus local Shimura varieties are determined by a purely group-theoretic datum without any underlying deformation problem. This is now an unpublished theorem, by the work of Fargues, Kedlaya-Liu, and Caraiani-Scholze. Cohomology of the local Shimura variety conjecturally provides a way to reconstruct the representation $r_\mu \circ LLC(\pi)$, where r_μ is the highest weight representation of $^L G$ associated naturally with the cocharacter μ of G (for simplicity, we suppose here that μ is defined over $\mathbb{Q}_p$).

The approach to local Langlands correspondence via cohomology of Rapoport-Zink spaces is explained in Section 2.3. First, in Section 2.4.1, we recall basics on p-divisible groups, including a relation to divisible commutative formal groups. Then, in Section 2.4.2, we explain Grothendieck-Messing theory of deformations of p-divisible groups over PD-thickenings. In particular, we introduce an important invariant of a p-divisible group, Dieudonné crystal. Section 2.4.3 is devoted to a classification of p-divisible groups over a perfect field. Up to isogeny, they are classified by their rational Dieudonné crystal with the action of Frobenius. In Section 2.4.4, we introduce Rapoport-Zink rigid analytic spaces $\mathcal{M}_{\mathbb{X},\eta}$ over L as certain universal deformation spaces of p-divisible groups, and in Section 2.4.5 we introduce a period map from $\mathcal{M}_{\mathbb{X},\eta}$ to the analytification of the Grassmannian $\mathrm{Gr}(d, n)$ over L. These constructions are based on the Grothendieck-Messing theory. In the case of Lubin-Tate spaces the period map coincides with the Gross-Hopkins period map to the analytification of a projective space $\mathbb{P}^{n-1}$ over L. We take into our considerations a level structure in Section 2.4.6 obtaining a pro-space $\mathcal{M}_{\mathbb{X},\infty}$ and then discuss an action of $W_{\mathbb{Q}_p}$ and $\boldsymbol{GL}_n(\mathbb{Q}_p)$ on its cohomology in the context of Kottwitz conjecture and local Langlands correspondence. Finally, in Section 2.4.7 we discuss a possible generalization of this to an arbitrary reductive group G over $\mathbb{Q}_p$.

Let us note that the main problem in the above approach is that in order to have a local Shimura variety being a (pro-)rigid analytic space, one needs the cocharacter μ of G to be minuscule. Thus one considers only a small part of all representations r_μ of $^L G$, which does not allow one to reconstruct the conjectural L-parameter $LLC(\pi)$. To overcome this, one needs to introduce new geometric objects instead of rigid analytic spaces. A short overview of recent developments in this direction is given in Section 2.4.8.

New geometric objects needed to have local Shimura varieties in a general case are so-called diamonds. In order to define them, first one introduces perfectoid spaces. These are glued from affinoid perfectoid spaces, which, in turn, are associated with perfectoid rings.

A perfectoid ring is a topological ring R with certain special properties (with respect to the fixed prime p). One defines the set $X = \mathrm{Spa}(R)$ of all equivalence classes of continuous valuations $|\cdot|\colon R \to \Gamma \cup \{0\}$ that satisfy certain conditions, where Γ is a totally ordered abelian group. The set X comes equipped with a natural topology and two natural presheaves of rings $\mathcal{O}_X$ and $\mathcal{O}_X^+$ related to R and R^+, respectively. A theorem of Scholze claims that the conditions on R to be perfectoid imply that the presheaves $\mathcal{O}_X$ and $\mathcal{O}_X^+$ are, actually, sheaves. This allows one to glue affinoid perfectoid spaces $(X, \mathcal{O}_X, \mathcal{O}_X^+)$ together.

One has étale topology on perfectoid spaces and diamonds are defined as algebraic spaces with respect to the pro-étale topology on perfectoid spaces. In particular, one can show that any rigid analytic space is a diamond. However, there are diamonds which are neither rigid analytic spaces, nor perfectoid spaces.

We introduce basics of the relevant p-adic geometry in Section 2.5. In Section 2.5.1, we define Huber rings and Tate rings, which are topological rings of special types. For example, $\mathbb{Z}_p$ is a Huber ring and $\mathbb{Q}_p$ is a Tate ring. More generally, an example of a Tate ring is given by any topological field with a non-archimedean norm. One gives an explicit description of all continuous valuations on a Tate ring R in Section 2.5.2. Namely, any continuous valuation factors uniquely through a valuation on a non-archimedean field and continuous valuations on a non-archimedean field are in bijection with all valuations on its quotient field. This allows one to work effectively with $\mathrm{Spa}(R, R^+)$. Section 2.5.3 contains a short description of adic spaces, perfectoid rings, and affinoid perfectoid spaces.

Having introduced perfectoids, we are in position to develop local Shimura varieties further. A new hero to consider is the ring of Witt vectors $W(\mathcal{O}_C)$ of the ring of integers $\mathcal{O}_C$ in a certain field C, which is (non-canonically) isomorphic to the completion of the algebraic closure of $\mathbb{F}_p((t))$. One has a natural action of Frobenius φ on $W(\mathcal{O}_C)$. A Fargues-Fontaine curve is the quotient X_{FF} of an open (non-affinoid) subspace of $\mathrm{Spa}(W(\mathcal{O}_C))$ by the action of φ, which is well-defined as an adic space. Note that X_{FF} behaves like a smooth projective curve over an algebraically closed field, though regular functions on X_{FF} form $\mathbb{Q}_p$. The curve X_{FF} has a canonical $\mathbb{C}_p$-point, which we denote by ∞. The completed local ring at ∞ is Fontaine's ring of periods B_{dR}^+.

A crucial observation by Fargues and Fontaine is that there is a canonical bijection between the set of isomorphism classes of $\bar{\mathbb{F}}_p$-isocrystals of dimension n and the set of isomorphism classes of vector bundle of rank n on the curve X_{FF}. On the other hand, there is a canonical isomorphism between the Galois group of $\mathbb{Q}_p$ and the étale fundamental group of X_{FF}. Furthermore, Scholze and Weinstein proved the following. Let $\mathbb{X}$ be a p-divisible group over $\bar{\mathbb{F}}_p$ of height n and dimension d and let $\mathcal{E}$ be the corresponding vector bundle on X_{FF}. Then there is a canonical bijection between the set of $\mathbb{C}_p$-points on the Rapoport-Zink space $\mathcal{M}_{\mathbb{X},\infty}$ and the set of trivial subbundles $\mathcal{O}_{X_{FF}}^{\oplus n} \subset \mathcal{E}$ such that the quotient $\mathcal{E}/\mathcal{O}_{X_{FF}}^{\oplus n}$ is the direct image with respect to the embedding $\infty \hookrightarrow X_{FF}$ of a vector space over $\mathbb{C}_p$ of dimension d. Altogether, this provides a linking bridge between local Langlands correspondence for $\mathbb{Q}_p$ and global Langlands correspondence on the Fargues-Fontaine curve.

One can develop this further and generalize it to an arbitrary local Shimura datum $(G, b, \bar{\mu})$. In this case one considers G-torsors on X_{FF} generalizing vector bundles, which correspond bijectively to Frobenius-twisted conjugacy classes in $G(L)$. Instead of trivial subbundles, one considers modifications of type $\bar{\mu}$ from a trivial torsor to a given one. Whenever $\bar{\mu}$ is minuscule, these modifications are parametrized by a pro-object in the category of rigid analytic spaces. For arbitrary $\bar{\mu}$, they are parametrized by diamonds, which are local Shimura varieties defined in a general case.

The approaches to local Langlands correspondence via cohomology of Lubin-Tate spaces, Rapoport-Zink spaces, and local Shimura varieties are translated in this language to the action of Hecke correspondences. Namely, one expects a version of Satake equivalence in the context of the Fargues-Fontaine curve. Also, there exists a stack Bun_G of G-bundles on the Fargues-Fontaine curve X_{FF}, which is smooth locally a diamond, that is, an "analytic Artin stack." Satake equivalence allows one to define for each representation V of $^L G$ a perverse sheaf on the Hecke correspondence $\mathrm{Hecke}_{\bar{\mu}} \to \mathrm{Bun}_G \times (\mathrm{Bun}_G \times X_{FF}^{abs})$, where X_{FF}^{abs} is an absolute Fargue-Fontaine curve, parametrizing points on X_{FF}. The latter kernel defines in a usual way a functor T_V from the derived category of ℓ-adic étale sheaves on Bun_G to that on $\mathrm{Bun}_G \times X_{FF}^{abs}$. One expects that this functor preserves constructibility and also that its image is in the full subcategory formed by $W_{\mathbb{Q}_p}$-equivariant sheaves on Bun_G (recall the isomorphism $\pi_1^{\acute{e}t}(X_{FF}) \simeq \mathrm{Gal}(\bar{\mathbb{Q}}_p/\mathbb{Q}_p)$).

Recall that in the geometric Langlands correspondence constructible sheaves on Bun_G live on the automorphic side. In our case, one has an open embedding $j: [*/G(\mathbb{Q}_p)] \hookrightarrow \mathrm{Bun}_G$, whence a smooth representation π of $G(\mathbb{Q}_p)$ defines a constructible sheaf $\mathcal{F}_\pi$ on $[*/G(\mathbb{Q}_p)]$ and then the sheaf $j_! \mathcal{F}_\pi$ on Bun_G. Given an irreducible π as above, one detects the corresponding L-parameter $LLC(\pi)$ by considering various finite collections $(V_1, \ldots, V_n)$ of representations of $^L G$, invariant vectors and covectors in $V_1 \otimes \ldots \otimes V_n$, finite collections $(w_1, \ldots, w_n)$ of elements in $W_{\mathbb{Q}_p}$, and playing with functors T_{V_i} applied to $j_! \mathcal{F}_\pi$. This is in analogy with what happens in the geometric Langlands correspondence and expresses the idea that $LLC(\pi)$ is an "eigenvalue" of $j_! \mathcal{F}_\pi$ with respect to all Hecke correspondences.

We sketch this program in Section 2.6. The Fargues-Fontaine curve X_{FF} is introduced in Section 2.6.1, where we also state results on relations between the curve and arithmetic objects over $\mathbb{Q}_p$. Then in Section 2.6.2 we introduce a $\mathbb{C}_p$-point ∞ on X_{FF} and cite a result on the relation between p-divisible groups and modifications of vector bundles on the Fargues-Fontaine curve at ∞. We pass to a general Shimura datum in Section 2.6.3, where we also discuss the geometry of the Grassmannian Gr_G, which is an ind-diamond. Hecke correspondences are introduced in Section 2.6.4 and we also show there a relation between smooth representations of $G(\mathbb{Q}_p)$ and constructible sheaves on Bun_G. Finally, in Section 2.6.5 we explain briefly how to detect the L-parameter $LLC(\pi)$ of an irreducible smooth representation π of $G(\mathbb{Q}_p)$ by applying Hecke correspondences.

2.2   LOCAL LANGLANDS CORRESPONDENCE

Fix two different prime numbers p and ℓ. Recall that the *Weil group* $W_{\mathbb{Q}_p} \subset \mathrm{Gal}(\bar{\mathbb{Q}}_p/\mathbb{Q})$ of $\mathbb{Q}_p$ is the dense subgroup mapping to integral powers of the Frobenius automorphism in $\mathrm{Gal}(\bar{\mathbb{F}}_p/\mathbb{F}_p)$. In other words, we have a cartesian diagram:

$$
\begin{array}{ccc}
W_{\mathbb{Q}_p} & \hookrightarrow & \mathrm{Gal}(\bar{\mathbb{Q}}_p/\mathbb{Q}_p) \\
\downarrow & & \downarrow \\
\mathbb{Z} & \hookrightarrow & \hat{\mathbb{Z}} = \mathrm{Gal}(\bar{\mathbb{F}}_p/\mathbb{F}_p)
\end{array}
$$

Let G be a reductive group over $\mathbb{Q}_p$. Define $^L G^\circ$ to be the connected, reductive group over $\bar{\mathbb{Q}}_\ell$ with based root datum dual to that of $G_{\bar{\mathbb{Q}}_p}$. One can show that the Galois group $\mathrm{Gal}(\bar{\mathbb{Q}}_p/\mathbb{Q}_p)$ acts on the algebraic group $^L G^\circ$ over $\bar{\mathbb{Q}}_\ell$ [3, Sect. 1, Sect. 2]. This leads to the following definition:

Definition 2.1.   The *L-group* or *Langlands dual* of G is the semidirect product $^L G := {}^L G^\circ \rtimes W_{\mathbb{Q}_p}$.

The natural inclusion and projection maps yield an exact sequence

$$
1 \longrightarrow {}^L G^\circ \longrightarrow {}^L G \longrightarrow W_{\mathbb{Q}_p} \longrightarrow 1.
$$

Definition 2.2.   An *L-parameter* of G is a continuous group splitting $\phi \colon W_{\mathbb{Q}_p} \to {}^L G$ of the above exact sequence up to conjugation by elements of $^L G^\circ$.

Example 2.3.   Let $G = \boldsymbol{GL}_n$ over $\mathbb{Q}_p$. Then $^L G^\circ = \boldsymbol{GL}_n$ over $\bar{\mathbb{Q}}_\ell$ with the trivial action of $W_{\mathbb{Q}_p}$ and $^L G$ is the direct product $^L G^\circ \times W_{\mathbb{Q}_p}$. Hence, giving an L-parameter is the same as giving a homomorphism $W_{\mathbb{Q}_p} \to \boldsymbol{GL}_n$ up to conjugation by $\boldsymbol{GL}_n$, that is, an n-dimensional continuous representation of $W_{\mathbb{Q}_p}$ over $\bar{\mathbb{Q}}_\ell$. An analogous result holds for any split reductive group G over $\mathbb{Q}_p$.

Recall that a *smooth representation* of $G(\mathbb{Q}_p)$ is a (possibly, infinite-dimensional) representation of $G(\mathbb{Q}_p)$ over $\bar{\mathbb{Q}}_\ell$ such that the stabilizer of any vector is an open subgroup of $G(\mathbb{Q}_p)$ with respect to the p-adic topology. Langlands made the following conjecture relating such representations to L-parameters:

Conjecture 2.4 (Local Langlands Correspondence).   *There exists a canonical map of sets*

$$LLC : \{\text{irreducible smooth representations of } G(\mathbb{Q}_p)\}/_{\simeq} \longrightarrow \{\text{L-parameters of } G\}$$

with finite fibers and certain special properties dependent on G.

For $G = \boldsymbol{GL}_n$ the Langlands conjecture was proven by Harris-Taylor [25] and Henniart [27].

In the above definitions, it is possible to replace the Weil group by the full Galois group $\mathrm{Gal}(\bar{\mathbb{Q}}_p/\mathbb{Q}_p)$. The case $G = \boldsymbol{GL}_1 = \mathbb{G}_m$ demonstrates why one should restrict to the Weil group in the conjecture. Indeed, it is known from local class field theory that in this case, both sides of the correspondence are bijective to the set of all continuous characters of $\mathbb{Q}_p^\times \simeq p^{\mathbb{Z}} \times \mathbb{Z}_p^\times$ with values in $\bar{\mathbb{Q}}_\ell^\times$.

The goal of these lectures will be to describe a program to construct a map as in the Langlands conjecture. The construction is based on the cohomology of so-called *local Shimura varieties* and generalizations thereof.

2.3   APPROACH TO LLC VIA LUBIN-TATE SPACES

2.3.1   Lubin-Tate Spaces

A general reference for formal groups is [26]. We begin by recalling the following definition:

Definition 2.5.   Let R be a ring. A *one-dimensional formal group law* over R is a formal power series $F \in R[[S,T]]$ such that

1. $F(S,T) = S + T + \text{higher degree terms}$;
2. associativity: $F(S, F(T,U)) = F(F(S,T), U)$.

The formal group law is called *commutative* if $F(S,T) = F(T,S)$. A homomorphism of formal group laws from F to G is a series $f \in R[[T]]$ with no constant term such that there is an equality $G(f(S), f(T)) = f(F(S,T))$.

The above axioms imply that there exists a unique series $\iota \in R[[T]]$ such that $F(T, \iota(T)) = 0$. If R has no nonzero torsion nilpotent elements, then any one-dimensional formal group law over R is commutative.

A one-dimensional formal group law F over R defines a group functor on the category of R-algebras that sends an R-algebra A to the set of its nilpotent elements with the group structure induced naturally by F. This gives a fully faithful functor from formal group laws to group functors on the category of R-algebras. Objects in its essential image are called *one-dimensional formal groups*. Equivalently, a one-dimensional formal group is a group functor which is represented as a set functor by the one-dimensional formal Lie variety over R, that is, by the formal spectrum $\mathrm{Spf}(R[[T]])$. In what follows we will often not distinguish between formal group laws and formal groups.

The homomorphisms between two commutative formal group laws form an abelian group. In particular, each integer $r \in \mathbb{Z}$ defines an endomorphism $[r] \in R[[T]]$ of a commutative formal group law F over R. Explicitly, if r is a natural number, then $[r](T) = F(\ldots F(F(T,T),T)\ldots)$, where the series F is applied $r-1$ times.

Let F be a (commutative) one-dimensional formal group law over a field of characteristic p. Then the derivative of $[p]$ vanishes and it follows that $[p]$ is a series in T^p. The *height* of the commutative formal group law is the largest natural number n such that $[p]$ is a series in T^{p^n}. Equivalently, it is the smallest natural number n such that the series $[p]$ has a nonzero coefficient by T^{p^n}. For example, the height of the multiplicative formal group law $S+T+ST$ is 1, because $[p] = T^p$ and, by definition, the height of the additive formal group law $S+T$ is infinite, because $[p] = 0$.

Fix a natural number n and let $\mathbb{X}$ be a (commutative) one-dimensional formal group over $\bar{\mathbb{F}}_p$ of dimension 1 and height n. It is known that $\mathbb{X}$ is unique up to isomorphism [26, Thm. 19.4.1]. Explicitly, $\mathbb{X}$ can be constructed as follows. Let $\mathbb{Q}_p \subset K$ be an unramified extension of degree n (which is unique up to isomorphism). Let σ denote the Frobenius automorphism both for $\bar{\mathbb{F}}_p$ and K, and choose p as a uniformizer in $\mathcal{O}_K$.

Definition 2.6. A *Lubin-Tate series* $f_p \in \mathcal{O}_K[[T]]$ is any series with linear term pT that is congruent to T^{p^n} modulo p (for instance, $f_p = pT + T^{p^n}$).

Choose any Lubin-Tate series $f_p \in \mathcal{O}_K[[T]]$. Then the condition $[p] = f_p$ uniquely determines a corresponding *Lubin-Tate formal group law* $F^{LT} \in \mathcal{O}_K$ $[[S,T]]$. A general reference for Lubin-Tate group laws is [8, § VI.3]. Note that different choices of Lubin-Tate series lead to canonically isomorphic formal group laws over $\mathcal{O}_K$. Taking

$$\mathbb{X} := F^{LT} \pmod{p} \in \mathbb{F}_{p^n}[[S,T]] \subset \bar{\mathbb{F}}_p[[S,T]],$$

we obtain the desired one-dimensional formal group of height n over $\bar{\mathbb{F}}_p$.

Any element $a \in \mathcal{O}_K$ uniquely defines an endomorphism $[a] = aT + \ldots$ of the Lubin-Tate group law. Reducing modulo p, we obtain an action of $\mathcal{O}_K$ on $\mathbb{X}$. Besides, one has a canonically defined Frobenius endomorphism φ of $\mathbb{X}$. If the coefficients of F^{LT} are in $\mathbb{Z}_p$, then the coefficients of $\mathbb{X}$ are in $\mathbb{F}_p$ and $\varphi(T) = T^p$. Otherwise, the series T^p gives a morphism of formal groups $T^p \colon \mathbb{X} \to \sigma(\mathbb{X})$, where $\sigma(\mathbb{X})$ is obtained by applying the Frobenius automorphism σ to the coefficients of the formal group law that corresponds to $\mathbb{X}$. We then define φ to be the composition of T^p with the isomorphism $\sigma(\mathbb{X}) \xrightarrow{\sim} \mathbb{X}$ induced by the canonical isomorphism $\sigma(F^{LT}) \xrightarrow{\sim} F^{LT}$ (note that $\sigma(F^{LT})$ is the Lubin-Tate formal group law associated with the Lubin-Tate series $\sigma(f_p)$).

The endomorphism φ of $\mathbb{X}$ satisfies the equalities

$$\varphi^n = [p], \qquad \varphi[a] = [\sigma(a)]\varphi$$

for any $a \in \mathcal{O}_K$. The $\mathbb{Z}_p$-algebra generated by $\mathcal{O}_K$ and a formal element Φ that satisfies these two properties of φ (namely, $\Phi^n = p$ and $\Phi a = \sigma(a)\Phi$) is isomorphic to the maximal order $\mathcal{O}_D$ of the central division $\mathbb{Q}_p$-algebra D with invariant $\frac{1}{n} \in \mathbb{Q}/\mathbb{Z} \simeq \mathrm{Br}(\mathbb{Q}_p)$. We thus obtain a homomorphism $\lambda \colon \mathcal{O}_D \to \mathrm{End}(\mathbb{X})$.

Proposition 2.7. *The above homomorphism $\lambda \colon \mathcal{O}_D \to \mathrm{End}(\mathbb{X})$ is an isomorphism.*

Proof. Injectivity follows from an explicit description of ideals in $\mathcal{O}_D$ and the fact that $[p^m] \neq 0$ for any natural m. For surjectivity note that any endomorphism of $\mathbb{X}$ over $\bar{\mathbb{F}}_p$ commutes with $\varphi^n = [p]$. Therefore, every endomorphism of $\mathbb{X}$ is defined over $\mathbb{F}_{p^n}$. Let $f(T) = aT + \ldots$ be an endomorphism of $\mathbb{X}$ over $\mathbb{F}_{p^n}$ and let $\hat{a} \in \mathcal{O}_K$ be the Teichmüller representative of $a \in \mathbb{F}_{p^n}$. Then the derivative of the difference $f - [\hat{a}]$ in $\mathrm{End}(\mathbb{X})$ vanishes, whence $f - [\hat{a}]$ is a series in T^p. It follows that there exists an endomorphism g of $\mathbb{X}$ such that $f - [\hat{a}]$ is equal to the composition $g\varphi$. Replacing f by g and iterating, we obtain a series $b_0 + b_1\Phi + \ldots + b_i\Phi^i + \ldots$ with $b_i \in \mathcal{O}_K$ and $b_0 = \hat{a}$, such that f and $\lambda(b_0 + \ldots + b_i\Phi^i)$ are congruent modulo $T^{p^{i+1}}$ for each $i \geqslant 0$. Since $\Phi^n = p$, the series $b_0 + b_1\Phi + \ldots + b_i\Phi^i + \ldots$ converges to an element of $\mathcal{O}_D$, which is sent by λ to f. $\square$

Now let us consider the formal deformation problem for $\mathbb{X}$ over the ring of Witt vectors $\mathcal{O}_L := W(\bar{\mathbb{F}}_p)$. Namely, let $\mathcal{M}^{(0)}$ be a functor on the category of Artinian $\mathcal{O}_L$-algebras with p nilpotent that sends such an algebra A with Jacobson radical $J = \mathrm{rad}(A) \subset A$ to the set

$$\left\{ \begin{array}{c} \text{one-dimensional commutative formal group } X \text{ over } A \\ \text{with an isomorphism} \\ \alpha \colon X \times_{\mathrm{Spec}(A)} \mathrm{Spec}(A/J) \xrightarrow{\sim} \mathbb{X} \times_{\mathrm{Spec}(\bar{\mathbb{F}}_p)} \mathrm{Spec}(A/J) \end{array} \right\} \Big/ \simeq . \qquad (2.1)$$

The following representability result is due to Lubin and Tate [38] (see also [11, Prop. 4.2]):

Proposition 2.8. *The functor $\mathcal{M}^{(0)}$ is represented by the formal scheme $\mathrm{Spf}\big(\mathcal{O}_L[[u_1, \ldots, u_{n-1}]]\big)$.*

More explicitly, let F be a one-dimensional formal group law over $\bar{\mathbb{F}}_p$ that corresponds to $\mathbb{X}$. Then $\mathcal{M}^{(0)}(A)$ is the set of all one-dimensional commutative formal group laws G over A such that $G \equiv F \pmod{J}$ taken up to isomorphisms of formal group laws over A inducing the identity modulo J. With this description, the parameters $u_1, \ldots, u_{n-1}$ correspond to the coefficients in $[p] \in A[[T]]$ by $T^p, \ldots, T^{p^{n-1}}$ (which are automatically nilpotent elements of A), where $[p]$ is associated with the formal group law G.

In what follows, we denote both the functor and the formal scheme representing it by $\mathcal{M}^{(0)}$. We call $\mathcal{M}^{(0)}$, which is the formal deformation space of $\mathbb{X}$, the *Lubin-Tate space*. By construction, we have a universal formal group $\mathcal{X}^{(0)}$ over $\mathcal{M}^{(0)}$. Since $\mathcal{O}_D \simeq \text{End}(\mathbb{X})$, we see that the group $\mathcal{O}_D^\times$ acts naturally on $\mathcal{M}^{(0)}$ by composing an isomorphism α in formula (2.1) with elements of $\mathcal{O}_D^\times$. For example, an A-point (X, α) of $\mathcal{M}^{(0)}$ is fixed by an element $a \in \mathcal{O}_D^\times$ if and only if the automorphism $\alpha^{-1}[a]\alpha$ of $X \times_{\text{Spec}(A)} \text{Spec}(A/p)$ extends to an automorphism of X.

Example 2.9. Consider the case $n = 1$. The one-dimensional formal group law of height 1 over $\bar{\mathbb{F}}_p$ is the multiplicative formal group law, and one can show that it has no nontrivial deformations. Thus $\mathcal{M}^{(0)}$ consists of a single point. Further, $D = \mathbb{Q}_p$ and the action of $\mathcal{O}_D^\times = \mathbb{Z}_p^\times$ on $\mathcal{M}^{(0)}$ is trivial. The latter also corresponds to the fact that any automorphism of the multiplicative formal group law over $\bar{\mathbb{F}}_p$ extends (uniquely) to an automorphism of the multiplicative formal group law over A (cf. Proposition 2.7).

To make things more symmetric, we consider the deformation space of $\mathbb{X}$ up to quasi-isogenies. Recall that there are no nonzero morphisms between one-dimensional formal groups of different heights and that any nonzero morphism between one-dimensional formal groups of the same height is called an *isogeny*. *Quasi-isogenies* arise from formally inverting isogenies. We denote quasi-isogenies by dashed arrows. It is clear from the above description of $\text{End}(\mathbb{X})$ that it is enough to invert $[p]$ and that quasi-isogenies from $\mathbb{X}$ to itself form the $\mathbb{Q}_p$-algebra D.

Let $\mathcal{M}$ be the formal deformation space of $\mathbb{X}$ up to quasi-isogenies. Namely, $\mathcal{M}$ represents the functor on the category of Artinian $\mathcal{O}_L$-algebras with p nilpotent that sends such an algebra A to the set

$$
\left\{
\begin{array}{c}
\text{one-dimensional commutative formal group } X \text{ over } A \\
\text{with a quasi-isogeny} \\
\alpha \colon X \times_{\text{Spec}(A)} \text{Spec}(A/p) \dashrightarrow \mathbb{X} \times_{\text{Spec}(\bar{\mathbb{F}}_p)} \text{Spec}(A/p)
\end{array}
\right\} / \simeq . \qquad (2.2)
$$

Note that since A is Artinian, the functor would be the same if we replace A/p by A/J, because quasi-isogenies between formal groups are lifted uniquely from A/J to A/p, where, as above, J is the Jacobson radical of A (see the end of Section 2.4.3).

Recall that the *height* of an isogeny is the largest natural number r such that the isogeny is given by a series in T^{p^r}. The height of quasi-isogenies defines the homomorphism $D^\times \to \mathbb{Z}$, which coincides with valuation of the reduced norm. For example, the height of φ is 1 and the height of $[p]$ is n. There is a decomposition

$$
\mathcal{M} = \coprod_{i \in \mathbb{Z}} \mathcal{M}^{(i)}
$$

according to the height of the quasi-isogeny α. The height zero component coincides indeed with $\mathcal{M}^{(0)}$ defined in formula (2.1), because quasi-isogenies of height zero from $\mathbb{X}$ to itself are just isomorphisms, being elements of the group $\mathcal{O}_D^\times$.

Now $\mathcal{M}$ is endowed with an action of $D^\times$, the group of quasi-isogenies from $\mathbb{X}$ to itself. Namely, a quasi-isogeny $\beta\colon \mathbb{X} \dashrightarrow \mathbb{X}$ sends a pair $(X,\alpha) \in \mathcal{M}(A)$ as in (2.2) to the pair $(X, \beta_{A/p} \circ \alpha) \in \mathcal{M}(A)$. The formal schemes $\mathcal{M}^{(i)}$ are isomorphic and the element $\Phi \in D^\times$ acts as a shift on $\mathcal{M}$ providing isomorphisms $\mathcal{M}^{(i)} \xrightarrow{\sim} \mathcal{M}^{(i+1)}$. Furthermore, there is a universal formal group $\mathcal{X}$ over $\mathcal{M}$.

Put $L := W(\bar{\mathbb{F}}_p)[\frac{1}{p}]$. Let $\mathcal{M}_\eta$ be the rigid analytic space over L which is the rigid analytic generic fiber of the formal scheme $\mathcal{M}$. Explicitly, $\mathcal{M}_\eta$ is isomorphic to the disjoint union over $\mathbb{Z}$ of open $(n-1)$-dimensional rigid analytic balls. By definition, there are equalities $\mathcal{M}_\eta(L) = \mathcal{M}(\mathcal{O}_L)$ and $\mathcal{M}_\eta(\mathbb{C}_p) = \mathcal{M}(\mathcal{O}_{\mathbb{C}_p})$. Furthermore, the universal formal group $\mathcal{X}$ induces a rigid analytic formal group $\mathcal{X}_\eta$ over $\mathcal{M}_\eta$. Because of characteristic zero, the p^m-torsion points of $\mathcal{X}_\eta$ define an étale locally constant sheaf $\mathcal{X}_\eta[p^m] \to \mathcal{M}_\eta$, which is étale locally isomorphic to $(\mathbb{Z}/p^m\mathbb{Z})^{\oplus n}$. This defines a $\boldsymbol{GL}_n(\mathbb{Z}/p^m\mathbb{Z})$-torsor $\mathcal{M}_m$ over $\mathcal{M}_\eta$ of framings of $X_\eta[p^m]$.

By definition, $\mathcal{M}_m$ classifies isomorphisms $\mathcal{X}_\eta[p^m] \simeq (\mathbb{Z}/p^m\mathbb{Z})^{\oplus n}$. Taken together, the $\mathcal{M}_m$ for $m \geqslant 1$ form a pro-system of étale covers of $\mathcal{M}_\eta$. Let $\mathcal{M}_\infty$ be the corresponding pro-object in the category of rigid analytic spaces (alternatively, one can associate a preperfectoid space with this pro-system [50, Thm. 6.3.4]). We see that $\mathcal{M}_\infty \to \mathcal{M}_\eta$ is a $\boldsymbol{GL}_n(\mathbb{Z}_p)$-torsor.

The action of $D^\times$ on $\mathcal{M}$ defines an action of $D^\times$ on $\mathcal{M}_\eta$, which extends naturally to an action of $D^\times$ on each $\mathcal{M}_m$ and on $\mathcal{M}_\infty$.

The above constructions are functorial with respect to the formal group $\mathbb{X}$. More precisely, given two one-dimensional formal groups $\mathbb{X}$ and $\mathbb{X}'$ over $\bar{\mathbb{F}}_p$ and an isogeny $\beta\colon \mathbb{X} \to \mathbb{X}'$, we obtain canonical morphisms $\mathcal{M} \to \mathcal{M}'$, $\mathcal{M}_m \to \mathcal{M}'_m$, and $\mathcal{M}_\infty \to \mathcal{M}'_\infty$ by composing α as in formula (2.1) with β. We denote these morphisms by β as well. By construction, β sends $\mathcal{M}_m^{(i)}$ to $(\mathcal{M}')_m^{(i+h)}$, where h is the height of the isogeny β.

The above observation has the following important consequence: though the rigid analytic spaces $\mathcal{M}_m$ over L are not defined canonically over $\mathbb{Q}_p$, they admit canonical descent data with respect to the extension $\mathbb{Q}_p \subset L$. Namely, there is a canonical isomorphism $F\colon \mathcal{M} \xrightarrow{\sim} \mathcal{M} \times_{L,\sigma} L$. Here $\times_{L,\sigma}$ denotes the pullback with respect to the Frobenius automorphism σ of L. This comes from the natural Frobenius isogeny $F\colon \mathbb{X} \to \mathbb{X} \times_{\bar{\mathbb{F}}_p, \sigma} \bar{\mathbb{F}}_p$ given by the series T^p and is composed of isomorphisms $\mathcal{M}_m^{(i)} \xrightarrow{\sim} \mathcal{M}_m^{(i+1)} \times_{L,\sigma} L$. Because of the shift, this descent data is non-effective. Otherwise, the Frobenius would not change infinitely many connected components of the rigid analytic space $\mathcal{M}_m$.

Another important fact about $\mathcal{M}_\infty$ is that the natural action of $\boldsymbol{GL}_n(\mathbb{Z}_p)$ extends to an action of $\boldsymbol{GL}_n(\mathbb{Q}_p)$ on the pro-object $\mathcal{M}_\infty$.[2] We explain this action

---

[2]The second action does not commute with the morphism to $\mathcal{M}_\eta$.

on the level of points $\mathcal{M}_\infty(\mathbb{C}_p)$. By construction, any such point x corresponds to the data (X, α, i) for a formal group X over $\mathcal{O}_{\mathbb{C}_p}$, a quasi-isogeny

$$\alpha : X \times_{\mathrm{Spec}(\mathcal{O}_{\mathbb{C}_p})} \mathrm{Spec}(\mathcal{O}_{\mathbb{C}_p}/p) \dashrightarrow \mathbb{X} \times_{\mathrm{Spec}(\bar{\mathbb{F}}_p)} \mathrm{Spec}(\mathcal{O}_{\mathbb{C}_p}/p) \,,$$

and an isomorphism

$$i : T_p X \xrightarrow{\sim} \mathbb{Z}_p^{\oplus n} \,,$$

where $T_p X := \varprojlim_m X[p^m](\mathbb{C}_p)$ is the *Tate module* of X. Let us define the action of a matrix $M \in \mathbf{GL}_n(\mathbb{Q}_p)$ on the point x. Applying the matrix M^{-1}, we obtain a new $\mathbb{Z}_p$-lattice $\Lambda := i^{-1}(M^{-1} \cdot \mathbb{Z}_p^{\oplus n})$ in the *rational Tate module* $V_p X := \mathbb{Q}_p \otimes_{\mathbb{Z}_p} T_p X$.

Claim 2.10. *There is a unique formal group X' over $\mathcal{O}_{\mathbb{C}_p}$ together with a quasi-isogeny $\beta \colon X \dashrightarrow X'$ such that $\beta(\Lambda) = T_p X'$, where we denote also by β the induced isomorphism between the rational Tate modules $\beta \colon V_p X \xrightarrow{\sim} V_p X'$.*

Proof. Since all $\mathbb{Z}_p$-lattices in $\mathbb{Q}_p^{\oplus n}$ are commensurable, it is enough to consider the case when one has an embedding $T_p X \subset \Lambda$. Then the kernel of the map $T_p X / p^m \longrightarrow \Lambda / p^m$ induced by the embedding $T_p X \subset \Lambda$ stabilizes for sufficiently large m. This kernel gives a subgroup

$$A \subset X[p^m](\mathbb{C}_p) \simeq T_p X / p^m \,,$$

which is independent of m whenever m is sufficiently large. The subgroup A extends uniquely to a finite flat subgroup $\tilde{A}$ of the finite flat group scheme $X[p^m]$ over $\mathcal{O}_{\mathbb{C}_p}$, which yields a formal group $X' := X/\tilde{A}$ over $\mathcal{O}_{\mathbb{C}_p}$ and an isogeny $\beta \colon X \to X'$. $\square$

By Claim 2.10, we have a quasi-isogeny α' defined by composing the isogenies

$$X' \times_{\mathrm{Spec}(\mathcal{O}_{\mathbb{C}_p})} \mathrm{Spec}(\mathcal{O}_{\mathbb{C}_p}/p) \xdashrightarrow{\beta^{-1}} X \times_{\mathrm{Spec}(\mathcal{O}_{\mathbb{C}_p})} \mathrm{Spec}(\mathcal{O}_{\mathbb{C}_p}/p) \xdashrightarrow{\alpha} \mathbb{X}$$

$$\times_{\mathrm{Spec}(\bar{\mathbb{F}}_p)} \mathrm{Spec}(\mathcal{O}_{\mathbb{C}_p}/p)$$

and an isomorphism i' defined by composing the isomorphisms

$$T_p X' \xrightarrow{\beta^{-1}} \Lambda \xrightarrow{i} M^{-1} \cdot \mathbb{Z}_p^{\oplus n} \xrightarrow{M} \mathbb{Z}_p^{\oplus n} \,.$$

Now we define $M(x) \in \mathcal{M}_\infty(\mathbb{C}_p)$ to be the point corresponding to the data (X', α', i'). This action of $\mathbf{GL}_n(\mathbb{Q}_p)$ extends $\mathbf{GL}_n(\mathbb{Z}_p)$-torsor structure on $\mathcal{M}_\infty$.

Note that for arbitrary m, the group $\boldsymbol{GL}_n(\mathbb{Q}_p)$ does not act on $\mathcal{M}_m$ itself, but on $\mathcal{M}_\infty$. One can check directly that a matrix $M \in \boldsymbol{GL}_n(\mathbb{Q}_p)$ sends $\mathcal{M}_\infty^{(i)}$ to $\mathcal{M}_\infty^{(i-r)}$, where $r \in \mathbb{Z}$ is the valuation of $\det(M) \in \mathbb{Q}_p^*$.

It is not hard to see that the actions of $D^\times$, the Frobenius F, and $\boldsymbol{GL}_n(\mathbb{Q}_p)$ on the rigid analytic space $\mathcal{M}_\infty$ commute with each other.

Example 2.11. Consider the case $n = 1$. Then there are canonical isomorphisms
$$\mathcal{M}^{(i)} \simeq \mathrm{Spec}(\mathcal{O}_L), \qquad \mathcal{M}_m^{(i)} \simeq \mathrm{Spec}\big(L(\sqrt[p^m]{1})\big).$$

Let us further identify $\mathcal{M}_m^{(i)}(\mathbb{C}_p)$ with $\mu_{p^m} \times \{p^{-i}\}$ (notice the inverse of the sign!). This defines isomorphisms
$$\mathcal{M}_m(\mathbb{C}_p) \simeq \mathbb{Q}_p(1)^\times / (1 + p^m \mathbb{Z}_p), \qquad \mathcal{M}_\infty(\mathbb{C}_p) \simeq \mathbb{Q}_p(1)^\times,$$

where $\mathbb{Q}_p(1)^\times$ denotes the set of all nonzero elements in $\mathbb{Q}_p(1) := \mathbb{Q}_p \otimes_{\mathbb{Z}_p} \varprojlim_m \mu_m$.
Note that $\mathbb{Q}_p(1)^\times$ is a $\mathbb{Q}_p^\times$-torsor. The action of $D^\times \simeq \mathbb{Q}_p^\times$ on $\mathbb{Q}_p(1)^\times$ is the inverse of the natural one, while the action of the group $\boldsymbol{GL}_1(\mathbb{Q}_p) \simeq \mathbb{Q}_p^\times$ is the natural one.

Note that, in general, the diagonally embedded subgroup $\mathbb{Q}_p^\times \subset D^\times \times \boldsymbol{GL}_n(\mathbb{Q}_p)$ acts trivially on $\mathcal{M}_\infty$.

2.3.2   Cohomology of Lubin-Tate Spaces and LLC

Now consider the *cohomology with compact support* of $\mathcal{M}_\infty$
$$H_c^*(\mathcal{M}_\infty) := \varinjlim_m H_{c,\text{ét}}^*(\mathcal{M}_{m,\mathbb{C}_p}, \bar{\mathbb{Q}}_\ell).$$

Since the $\mathcal{M}_m$ are defined over L, we obtain a canonical continuous action of the Galois group $\mathrm{Gal}(\bar{L}/L) \simeq \mathrm{Gal}(\bar{\mathbb{Q}}_p/\mathbb{Q}_p^{\mathrm{nr}})$ on this cohomology. The isomorphism $F \colon \mathcal{M} \xrightarrow{\sim} \mathcal{M} \times_{L,\sigma} L$ allows us to extend this action to a continuous action of the Weil group $W_{\mathbb{Q}_p}$.

Thus we obtain an action of the group $D^\times \times W_{\mathbb{Q}_p} \times \boldsymbol{GL}_n(\mathbb{Q}_p)$ on $H_c^*(\mathcal{M}_\infty, \bar{\mathbb{Q}}_\ell)$. The restriction of this action to the subgroup $W_{\mathbb{Q}_p}$ is continuous. The restrictions of this action to the subgroups $D^\times$ and $\boldsymbol{GL}_n(\mathbb{Q}_p)$ are smooth representations of these groups. Indeed, the open subgroup $\mathrm{Ker}\big(\boldsymbol{GL}_n(\mathbb{Z}_p) \to \boldsymbol{GL}_n(\mathbb{Z}/p^m\mathbb{Z})\big)$ of $\boldsymbol{GL}_n(\mathbb{Q}_p)$ acts identically on $\mathcal{M}_m$ and hence stabilizes $H_{c,\text{ét}}^*(\mathcal{M}_{m,\mathbb{C}_p}, \bar{\mathbb{Q}}_\ell)$. That the action of $D^\times$ is smooth follows from the fact that for each level m the action of $D^\times$ on $\mathcal{M}_m$ is continuous [18, Prop. 2.3.11]. The representation $H_c^{n-1}(\mathcal{M}_\infty)$ of the group $D^\times \times W_{\mathbb{Q}_p} \times \boldsymbol{GL}_n(\mathbb{Q}_p)$ is the main linking bridge between representations of $\boldsymbol{GL}_n(\mathbb{Q}_p)$ and $W_{\mathbb{Q}_p}$.

Example 2.12. Consider the case $n=1$. By Example 2.11, we see that H_c^0 $(\mathcal{M}_\infty)$ is isomorphic to $\bar{\mathbb{Q}}_l[\mathbb{Q}_p(1)^\times]$, the "group algebra" of the $\mathbb{Q}_p^\times$-torsor $\mathbb{Q}_p(1)^\times$. The action of $D^\times \simeq \mathbb{Q}_p^\times$ on $\bar{\mathbb{Q}}_l[\mathbb{Q}_p(1)^\times]$ is the inverse of the natural action, while the action of $\boldsymbol{GL}_1(\mathbb{Q}_p) \simeq \mathbb{Q}_p^\times$ is the natural one and the action of the group $W_{\mathbb{Q}_p}$ is through the quotient $W_{\mathbb{Q}_p}^{\mathrm{ab}} \simeq \mathbb{Q}_p^\times$ (the latter isomorphism is from local class field theory).

It follows that for any continuous $\bar{\mathbb{Q}}_l$-valued character χ (that is, for any irreducible smooth representation) of $\boldsymbol{GL}_1(\mathbb{Q}_p) \simeq \mathbb{Q}_p^\times$, we have that $\mathrm{Hom}_{\boldsymbol{GL}_1(\mathbb{Q}_p)}$ $(\chi, H_c^0(\mathcal{M}_\infty))$ is a one-dimensional space endowed with the action of $D^\times \times W_{\mathbb{Q}_p}$ given by $\chi^{-1} \otimes \chi$.

There is a method for constructing representations of $\boldsymbol{GL}_n$ called *parabolic induction*. Namely, given a parabolic subgroup $P \subset \boldsymbol{GL}_n$ with the corresponding semisimple Levi quotient M (which is necessarily isomorphic to $\boldsymbol{GL}_{n_1} \times \ldots \times \boldsymbol{GL}_{n_r}$ with $n_1 + \ldots + n_r = n$) and a smooth representation ρ of M, the parabolic induction of ρ is the induction with compact support $\mathrm{ind}_M^G(\rho)$, where we denote also by ρ the restriction of ρ from M to P.

Definition 2.13. A *supercuspidal* representation of $\boldsymbol{GL}_n(\mathbb{Q}_p)$ is an irreducible smooth representation which is not isomorphic to a subrepresentation (equivalently, quotient) of any parabolically induced representation.

Let π be an (irreducible) supercuspidal representation of $\boldsymbol{GL}_n(\mathbb{Q}_p)$ over $\bar{\mathbb{Q}}_\ell$ and consider the representation of $D^\times \times W_{\mathbb{Q}_p}$ on the space

$$H_c^*(\mathcal{M}_\infty)[\pi] := \mathrm{Hom}_{\boldsymbol{GL}_n(\mathbb{Q}_p)}\big(\pi, H_c^*(\mathcal{M}_\infty)\big).$$

Recall that every supercuspidal representation is a discrete series representation, for which we have the *Jacquet-Langlands correspondence* (see [33] for $\boldsymbol{GL}_2$ and [46], [9] for the general case). This gives a bijection

$$JL : \{\text{irreducible discrete series representations of } \boldsymbol{GL}_n(\mathbb{Q}_p)\}/_\simeq \overset{1:1}{\longleftrightarrow}$$
$$\{\text{irreducible representations of } D^\times\}/_\simeq.$$

For supercuspidal representations we also have the *local Langlands correspondence*, which gives a bijection

$$LLC : \{\text{supercuspidal representations of } \boldsymbol{GL}_n(\mathbb{Q}_p)\}/_\simeq \overset{1:1}{\longleftrightarrow}$$
$$\{\text{irreducible } n\text{-dimensional representations of } W(\mathbb{Q}_p)\}/_\simeq.$$

We have the following theorem of Harris and Taylor [25], which was previously conjectured by Carayol [7]:

Theorem 2.14. *For any supercuspidal representation π of $\boldsymbol{GL}_n(\mathbb{Q}_p)$ over $\bar{\mathbb{Q}}_\ell$, up to duals and twists, the following tensor product decomposition holds:*

$$H_c^{n-1}(\mathcal{M}_\infty)[\pi] \simeq JL(\pi) \otimes LLC(\pi).$$

Remark 2.15. This is a key step in the proof of (the existence of) LLC. By results of Faltings, Strauch, and others, one can use a Lefschetz trace formula argument to show, using only local methods, that

$$H_c^{n-1}(\mathcal{M}_\infty)[\pi] = JL(\pi) \otimes V,$$

where V is some n-dimensional representation of $W_{\mathbb{Q}_p}$. One can declare $LLC(\pi) := V$. However, this does not prove local Langlands for $\boldsymbol{GL}_n$. One needs to show many more properties (irreducibility of V, bijectivity of LLC...) that can only be shown using global arguments.

Our goal is to generalize the above associations to a map

$$\{\text{irreducible smooth representations of } G(\mathbb{Q}_p)\}/_\simeq \longrightarrow \{L\text{-parameters of } G\}$$

as predicted by the Langlands conjecture. We first generalize the Lubin-Tate spaces to what are known as *Rapoport-Zink spaces*. These are moduli spaces of p-divisible groups with extra structure and exist for $\boldsymbol{GL}_n$, $\boldsymbol{GSp}_{2n}$, $\boldsymbol{GU}_n$, and $\boldsymbol{GO}_n$. In the next section, we explain the case of $\boldsymbol{GL}_n$.

2.4 APPROACH TO LLC VIA RAPOPORT-ZINK SPACES

2.4.1 Reminders on p-divisible Groups

In order to proceed, we have to enlarge the category of objects under consideration from one-dimensional commutative formal groups to p-divisible groups. When doing this, we restrict ourselves to divisible (or, equivalently, finite height) formal groups only. Lubin-Tate formal groups are divisible and do correspond to p-divisible groups but, for instance, the additive formal group is not so in positive characteristic. Hence, the restriction to divisible formal groups here excludes some formal groups but luckily not those of importance to us.

The notion of p-divisible groups was introduced by Serre and Tate in their study of finite torsion group schemes associated with abelian varieties. Standard references are the original exposition of Tate [51] and Messing's book [41], and also a survey can be found in [47].

Let S be a scheme on which p is locally nilpotent.

Definition 2.16. A *p-divisible group* over S is an fpqc sheaf (of abelian groups) X over S of the form $X = \bigcup_{m \geqslant 1} X[p^m]$, where each $X[p^m]$ is a p^m-torsion finite locally free commutative group scheme over S, and for all m, m', we have an exact sequence of fpqc sheaves

$$0 \longrightarrow X[p^m] \longrightarrow X[p^{m+m'}] \overset{p^m}{\longrightarrow} X[p^{m'}] \longrightarrow 0. \tag{2.3}$$

In other words, $X[p^m]$ is the kernel of the faithfully flat morphism $p^m \colon X[p^{m+m'}] \to X[p^{m'}]$. It follows immediately from exact sequences (2.3) that there exists a number n such that the structure sheaf of each $X[p^m]$ is of rank p^{mn} over S. This number n is called the *height* $\mathrm{ht}(X)$ of X.

One defines $\mathrm{Lie}(X)$ as the group functor that sends an S-scheme T to

$$\mathrm{Ker}\big(X(T[\varepsilon]) \to X(T)\big),$$

where ε is a formal variable that satisfies $\varepsilon^2 = 0$ and $T[\varepsilon] := \mathrm{Spec}_T(\mathcal{O}_T[\varepsilon])$. Using that p is locally nilpotent on S, one can show that $\mathrm{Lie}(X)$ is a vector bundle on S (for example, this follows from [41, Ch. II, Thm. 3.3.13]). The *dimension* $\dim(X)$ of the p-divisible group X is defined to be the rank of the vector bundle $\mathrm{Lie}(X)$ over S. There is an inequality

$$\dim(X) \leqslant \mathrm{ht}(X).$$

For any p-divisible group X, the Cartier duals $X[p^m]^\vee := \mathrm{Hom}(X[p^m], \mathbb{G}_m)$ assemble to another p-divisible group $X^\vee$, which is called the *Serre dual* of X. One has the following relation [51, Prop. 3]:

$$\dim(X) + \dim(X^\vee) = \mathrm{ht}(X).$$

Here are some examples of p-divisible groups:

Example 2.17.

(i) $X = \mathbb{Q}_p/\mathbb{Z}_p = \bigcup_{m \geqslant 1} \frac{1}{p^m}\mathbb{Z}/\mathbb{Z}$; we have $\mathrm{ht}(X) = 1$ and $\dim(X) = 0$.

(ii) $X = \mu_{p^\infty} = \bigcup_{m \geqslant 1} \mu_{p^m}$; we have $\mathrm{ht}(X) = 1$ and $\dim(X) = 1$; μ_{p^∞} and $\mathbb{Q}_p/\mathbb{Z}_p$ are Serre dual to each other.

(iii) $X = A[p^\infty] = \bigcup_{m \geqslant 1} A[p^m]$ where A is an abelian scheme over S of relative dimension g; we have $\mathrm{ht}(X) = 2g$ and $\dim(X) = g$; $A[p^\infty]$ and $A^\vee[p^\infty]$ are Serre dual to each other, where $A^\vee$ is the dual abelian scheme.

Let us also discuss briefly a relation between formal groups and p-divisible groups. Assume that $S = \mathrm{Spec}(R)$, where R is a complete Noetherian local ring such that the characteristic of the residue field is p. Let $F \in R[[S_1, \ldots, S_d, T_1, \ldots, T_d]]$ be a commutative formal group law of dimension d such that F is divisible. This means that for each $m \geqslant 1$, we have a finite locally free commutative group scheme $F[p^m]$ over S whose ring of regular functions is $R[[T_1, \ldots, T_d]]/$

([p^m]), where [p^m] is the collection of d series that corresponds to the endomorphism [p^m] of F. The group schemes $F[p^m]$, $m \geqslant 1$, form a p-divisible group, which is *connected*, that is, the rings of regular functions on $F[p^m]$ are local. This defines an equivalence between the category of divisible commutative formal groups over S and the category of connected p-divisible groups over S (see [51, §2.2]).

It is worth mentioning in this context that any p-divisible group X over S fits into a uniquely defined exact sequence

$$0 \longrightarrow X^0 \longrightarrow X \longrightarrow X^{\acute{e}t} \longrightarrow 0\,,$$

where X^0 is a connected p-divisible group and $X^{\acute{e}t}$ is an étale p-divisible group [51, §1.4].

One finds in [41, Ch. II] a detailed treatment of relations between formal groups and p-divisible groups in the general case when S is an arbitrary scheme on which p is locally nilpotent.

2.4.2 Grothendieck-Messing Deformation Theory

Grothendieck-Messing theory classifies possible extensions X' over S' of a p-divisible group X over S as above along a PD-thickening $S \hookrightarrow S'$. Naturally, such extensions can be considered as deformations of X over S'. The original references for Grothendieck-Messing theory are [23], [40], and [41]. Short surveys are given in Grothendieck's ICM talk [24] and in [5, Sect. 12.1].

Of fundamental importance for the theory is the existence of a *universal vector extension* [41, Ch. IV]. This is an exact sequence of fpqc sheaves on S

$$0 \longrightarrow V \longrightarrow EX \longrightarrow X \longrightarrow 0\,,$$

where V is a vector bundle on S considered a sheaf of abelian groups, such that any other extension of this type arises by a unique push-out. It can be seen that V is isomorphic to the sheaf $\omega_{X^\vee}$ of differential forms on the Serre dual $X^\vee$. In other words, $\omega_{X^\vee}$ is the dual of the Lie vector bundle $\mathrm{Lie}(X^\vee)$.

Examples 2.18. We continue Example 2.17.

(i) If $X = \mathbb{Q}_p/\mathbb{Z}_p$, then $EX = (\mathbb{G}_a \oplus \mathbb{Q}_p)/\mathbb{Z}_p$ where $\mathbb{Z}_p$ acts diagonally (since p is locally nilpotent on S, we have a natural morphism $\mathbb{Z}_p \to \mathbb{G}_a$).
(ii) If $X = \mu_{p^\infty}$, then $EX = \mu_{p^\infty}$.
(iii) If $X = A[p^\infty]$, then $EX = (EA)[p^\infty]$, where EA is the universal vector extension of A. Recall that if A is an abelian variety over $\mathbb{C}$, then $EA \simeq H_1(A, \mathbb{C})/H_1(A, \mathbb{Z})$ and

$$V \simeq H^0(A, \Omega^1_{A^\vee}) \simeq H^1(A, \mathcal{O}_A)^\vee \simeq F^{-1} H_1(A, \mathbb{C})$$

is the only nontrivial term of the Hodge filtration on $H_1(A, \mathbb{C}) \simeq H^1(A, \mathbb{C})^\vee$.

The next theorem is due to Grothendieck and Messing [41, Ch. IV, V][3] and shows the crystalline nature of the universal vector extension EX. To be precise, the first part shows that it gives rise to a crystal $M(X)$ ("something that grows in a rigid manner," or more down-to-earth: a sheaf on the crystalline site of S) by taking its Lie algebra. In fact, a similar assertion can be made about the universal extension itself considered as an fpqc sheaf.

Before stating the theorem, recall that a *PD-thickening* of S is a nilpotent closed immersion $S \hookrightarrow S'$ such that the ideal $I := \mathrm{Ker}(\mathcal{O}_{S'} \to \mathcal{O}_S)$ has a nilpotent PD-structure. The latter means that we have maps, called *divided powers*, from I to itself $x \mapsto \gamma_i(x)$ "of kind $x \mapsto \frac{x^i}{i!}$" such that $\gamma_{i_1}(I) \cdot \ldots \cdot \gamma_{i_r}(I) = 0$ whenever $i_1 + \ldots + i_r$ is sufficiently large (see, e.g., [2] for details on divided powers).

By a deformation of X over S', we mean a pair (X', ϕ), where X' is a p-divisible group over S' and $\phi \colon X' \times_{S'} S \xrightarrow{\sim} X$ is an isomorphism of p-divisible groups over S.

Theorem 2.19. *Let S and X be as above and let $S \hookrightarrow S'$ be a PD-thickening. Then the following is true:*

(i) *The fpqc sheaf EX' on S' is canonically independent of the choice of a deformation (X', ϕ) of X over S'. Set $M(X)(S') := \mathrm{Lie}(EX')$, which is a vector bundle on S' of rank $\mathrm{ht}(X)$.*

(ii) *The set of deformations (X', ϕ) up to isomorphism corresponds bijectively to the set of vector subbundles $V' \subset M(X)(S')$ (in particular, the quotient $M(X)(S')/V'$ is locally free) such that $V' \otimes_{\mathcal{O}_{S'}} \mathcal{O}_S \simeq V \subset M(X)(S) = \mathrm{Lie}(EX)$.*

We call $M(X)$ the *covariant Dieudonné crystal.* Note that there are equalities

$$\mathrm{rk}(V') = \mathrm{ht}(X) - \dim(X) = \dim(X^\vee).$$

Theorem 2.19(*ii*) can be interpreted as saying that "deforming X is equivalent to deforming the Hodge filtration."

2.4.3   *p*-divisible Groups Over a Perfect Field

We now discuss the classification of p-divisible groups over $S = \mathrm{Spec}(k)$, where k is a perfect field of characteristic p. Put $\mathcal{O}_K := W(k)$, $K := W(k)[\frac{1}{p}]$, and denote the Frobenius on K by σ.

---

[3]As mentioned in [41], Messing's work depends on the (then unproven) fact that any p-divisible group X over S lifts to a p-divisible group X' over a nilpotent thickening S' of S. Otherwise, $M(X)(S')$ might be undefined for certain thickenings S'. This was proven by Grothendieck using the description of deformation obstructions for group schemes in Illusie's thesis [31]. His result is given as [32, Théorème 4.4].

For each m, we have that $\mathcal{O}_K/p^m \to k$ is a PD-thickening and $\mathcal{O}_K \to k$ is a topological PD-thickening (the latter means that a PD-structure is topologically nilpotent). The reason is that $\frac{p^i}{i!} \in p\mathcal{O}_K$ and $\frac{p^i}{i!} \to 0$ as $i \to \infty$ (provided that $p \neq 2$, but this can be easily overcome). Thus by Theorem 2.19(i), given a p-divisible group X over k, we obtain a well-defined free $\mathcal{O}_K$-module $M(X)(\mathcal{O}_K)$ of rank $\mathrm{ht}(X)$, which we denote just by $M(X)$.

See [10] and [20] for the following theorem of Dieudonné (see also a good overview on this and on material in the next section in [5, Ch. 7, Ch. 8]):

Theorem 2.20. *There is an equivalence of categories*

$$\{p\text{-divisible groups over } k\} \xrightarrow{\sim}$$

$$\left\{ \begin{array}{c} (M, F, V), \text{ where } M \text{ is a free finite rank } \mathcal{O}_K\text{-module,} \\ F: M \to M \text{ is } \sigma\text{-linear,} \\ V: M \to M \text{ is } \sigma^{-1}\text{-linear,} \\ \text{and } FV = VF = p \end{array} \right\}$$

that sends X to $(M(X), F, V)$, where F is induced by the Frobenius morphism $X \to X \times_{\sigma,k} k$ and V is induced by the Verschiebung $X \times_{\sigma,k} k \to X$ defined by the Frobenius morphism for the Serre dual p-divisible group $X^\vee$.

In notation of Theorem 2.20, V is defined uniquely by F. One sees that giving a triple (M, F, V) is the same as giving a pair (M, F) such that $F(M) \supset p \cdot M$.

Examples 2.21. We continue Example 2.17.

(i) If $X = \mathbb{Q}_p/\mathbb{Z}_p$, then $(M(X), F, V) = (\mathcal{O}_K, p\sigma, \sigma^{-1})$.
(ii) If $X = \mu_{p^\infty}$, then $(M(X), F, V) = (\mathcal{O}_K, \sigma, p\sigma^{-1})$.
(iii) If $X = A[p^\infty]$, then there are equalities [40, Thm. 1]

$$(M(X), F, V) = \left(H^1_{\mathrm{crys}}(A/\mathcal{O}_K)^\vee, V_A^\vee, F_A^\vee\right) = \left(H^1_{\mathrm{crys}}(A^\vee/\mathcal{O}_K), F_{A^\vee}, V_{A^\vee}\right).$$

The category of p-divisible groups over k up to isogeny is very simple to describe, which was originally noticed by Dieudonné and Manin. The name "isocrystal" was coined by Grothendieck and is evidently a concatenation of "iso"geny and "crystal" with the meaning "Dieudonné crystal up to isogeny."

Definition 2.22. A *k-isocrystal* is a pair (W, φ_W), where W is a finite-dimensional K-vector space and $\varphi_W: W \to W$ is a σ-linear isomorphism.

The following classification result is due to Dieudonné and Manin [39, Sect. 2.4], [10, Ch. 4]:

Theorem 2.23.

(i) *Suppose that k is algebraically closed. Then the category of k-isocrystals is semisimple and the set of isomorphism classes of simple objects is bijective with $\mathbb{Q}$. Explicitly, we associate to $\mathbb{Q} \ni \lambda = \frac{r}{s}$ with $(r, s) = 1$, $s > 0$, the isocrystal*

$$D_\lambda := (K^s, b_\lambda \circ \sigma),$$

where b_λ is the linear map $K^s \to K^s$ determined by the $(s \times s)$-matrix

$$b_\lambda = \begin{pmatrix} 0 & 1 & & & \\ & & \ddots & & \\ & & & \ddots & 1 \\ p^r & & & & 0 \end{pmatrix}$$

and σ denotes also the component-wise application of the Frobenius σ on K^s. For an arbitrary k-isocrystal, its slopes are λ's associated to its simple summands.

(ii) *When k is an arbitrary perfect field of characteristic p, define slopes of a k-isocrystal as slopes of its scalar extension to $W(\bar{k})[\frac{1}{p}]$. Then the category of p-divisible groups over k up to isogeny is equivalent to the category of k-isocrystals with slopes $\lambda \in [0, 1]$. The equivalence sends a p-divisible group X over k to the isocrystal $M(X)_K$ with the Frobenius F.*

Theorem 2.23 can be used to define the *Newton polygon of a p-divisible group*. In fact, we can attach a Newton polygon $\mathrm{NP}(W)$ to any isocrystal W over an algebraically closed field k in the following way. By Theorem 2.23, W is isomorphic to a (unique) direct sum $\bigoplus_\lambda D_\lambda^{\oplus m_\lambda}$. Let

$$\lambda_1 \leqslant \lambda_2 \leqslant \ldots \leqslant \lambda_n, \qquad n := \dim_K(W),$$

be the ascending sequence where each slope λ of W occurs exactly $\dim(D_\lambda) \cdot m_\lambda$ times. Then the Newton polygon $\mathrm{NP}(W)$ of W is the polygon with vertices

$$P_i = (i, \lambda_1 + \ldots + \lambda_i), \qquad 0 \leqslant i \leqslant n,$$

and edges $P_i P_{i+1}$, $0 \leqslant i \leqslant n - 1$. By definition, the breakpoints of $\mathrm{NP}(W)$ are integral.

Now, the Newton polygon $\mathrm{NP}(X)$ of a p-divisible group X over k is just the Newton polygon of its isocrystal $M(X)_K$. By Theorem 2.23, $\mathrm{NP}(X)$ determines X up to isogeny and all slopes of $M(X)_K$ are contained in $[0, 1]$. Moreover, the last vertex has coordinates (n, d), where $n = \mathrm{ht}(X)$ and $d = \dim(X)$. A typical Newton polygon of a p-divisible group looks like this:

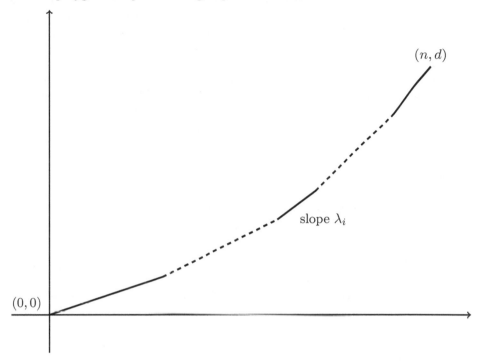

There is also a more basic interpretation of the Newton polygon explaining its name. Any $\bar{\mathbb{F}}_p$-isocrystal (W, φ_W) is the scalar extension of some $\mathbb{F}_q$-isocrystal (W_0, φ_{W_0}), where W_0 is a vector space over the field $K_0 := W(\mathbb{F}_q)[\frac{1}{p}]$ and $q = p^e$ for some e. Although φ_{W_0} is not K_0-linear, its e-th iteration $\varphi^e_{W_0} : W_0 \to W_0$ is and we may consider its characteristic polynomial

$$\chi(T) := \det(\varphi^e_{W_0} - T) \in K_0[T].$$

Up to rescaling, $\mathrm{NP}(W)$ is the (classical) Newton polygon of the polynomial $\chi(T)$. Namely, the collection $\lambda_1, \ldots, \lambda_n$ is the collection of p-adic valuations of the roots of $\chi(T)$, where the valuation is normalized so that the valuation of q is 1 (see [10, Sect. IV.6]).

One may also ask for a classification of isogeny classes of p-divisible groups over the thickening $\mathcal{O}_K/p^m$ of k. The answer remains the same because of Drinfeld rigidity [12], [1, Thm. II.2.2.3], [34, Lem. 1.1.3]: given two p-divisible groups G and G' over $\mathcal{O}_K/p^m$, every quasi-isogeny $G_k \to G'_k$ over k admits a unique lifting $G \to G'$ over $\mathcal{O}_K/p^m$. Consequently, the isogeny classes of p-divisible groups over k and over $\mathcal{O}_K/p^m$ correspond one-to-one to each other and there is nothing new to classify.

2.4.4 Definition of Rapoport-Zink Spaces and Basic Results

Having classified p-divisible groups over $W(\bar{\mathbb{F}}_p)/p^m$ up to isogeny, we now consider a more refined sort of moduli problem concerning p-divisible groups in single isogeny classes. This moduli problem gives rise to Rapoport-Zink spaces as introduced in [45]. We consider here only a small portion of their theory, ignoring the more versatile moduli problems involving additional structures such as polarizations and endomorphisms. In fact, our moduli problem $\mathcal{M}_{\mathbb{X}}$ below corresponds to the simplest case $G = \boldsymbol{GL}_n$ of their theory. The ICM talk of Rapoport [42] contains a brief survey on Rapoport-Zink spaces and their applications.

Let $\mathbb{X}$ be a p-divisible group over $\bar{\mathbb{F}}_p$ of height n and dimension d. Put $\mathcal{O}_L := W(\bar{\mathbb{F}}_p)$ and $L := W(\bar{\mathbb{F}}_p)$.

Define a functor $\mathcal{M}_{\mathbb{X}}$ on the category of $\mathcal{O}_L$-algebras with p being locally nilpotent that sends such an algebra A to the set

$$\left\{ \begin{array}{c} p\text{-divisible group } X \text{ over } S \text{ with a quasi-isogeny} \\ \rho\colon X \times_{\mathrm{Spec}(A)} \mathrm{Spec}(A/p) \dashrightarrow \mathbb{X} \times_{\mathrm{Spec}(\bar{\mathbb{F}}_p)} \mathrm{Spec}(A/p) \end{array} \right\} \Big/ \simeq .$$

In other words, the functor $\mathcal{M}_{\mathbb{X}}$ classifies liftings of p-divisible groups in the isogeny class of $\mathbb{X}$ modulo compatible isomorphisms.

Recall that a morphism $X \to Y$ of Noetherian formal schemes is called (locally) formally of finite type if $X^{\mathrm{red}} \to Y^{\mathrm{red}}$ is (locally) of finite type. In particular, for a formal scheme being locally formally of finite type over $\mathcal{O}_L$ means that it can be covered by affine formal schemes associated to quotients of rings having the form

$$\mathcal{O}_L[[X_1, \ldots, X_r]] \langle Y_1, \ldots, Y_s \rangle . \tag{2.4}$$

This adic ring is the completion of $\mathcal{O}_L[X_1, \ldots, X_r, Y_1, \ldots, Y_s]$ with respect to the adic topology generated by $(p, X_1, \ldots, X_r)$.

Theorem 2.24 (Rapoport-Zink [45, Thm. 2.16]). *The functor $\mathcal{M}_{\mathbb{X}}$ is representable by a formal scheme, which is locally formally of finite type over $\mathcal{O}_L$.*

As usual, we abuse notation and write $\mathcal{M}_{\mathbb{X}}$ also for the formal scheme representing $\mathcal{M}_{\mathbb{X}}$. We collect now some useful further properties of $\mathcal{M}_{\mathbb{X}}$.

Theorem 2.25.

(i) $\mathcal{M}_{\mathbb{X}}$ is formally smooth. (This just means existence of nilpotent deformations.)
(ii) $\mathcal{M}_{\mathbb{X}}$ has dimension $d \cdot (n - d)$.

The rigid analytic generic fiber $\mathcal{M}_{\mathbb{X},\eta}$ over L of the formal scheme $\mathcal{M}_{\mathbb{X}}$ [45, Ch. 5] is an instance of a *Rapoport-Zink space*.

Example 2.26. If $\mathbb{X}$ is the connected p-divisible group associated with the one-dimensional formal group of finite height n, then the restriction of $\mathcal{M}_{\mathbb{X}}$ to

Artinian rings over $\mathcal{O}_L$ coincides with the functor $\mathcal{M}$ defined in formula (2.2). The formal schemes obtained in Lubin-Tate theory coincide with those provided by Theorem 2.24. For general $\mathbb{X}$, we have to work with all $\mathcal{O}_L$-algebras with p being locally nilpotent instead of Artinian rings over $\mathcal{O}_L$. In particular, a restriction to Artinian rings would lead to $s = 0$ in (2.4).

2.4.5   Period Map

The Rapoport-Zink space $\mathcal{M}_{\mathbb{X},\eta}$ can be endowed with a period morphism by Grothendieck-Messing deformation theory. We indicate some details in this section and refer to [1, Ch. II] and [45, Ch. 5] for more details. The Grassmannian $\mathrm{Gr}(d,n)$ is the algebraic variety over L classifying d-dimensional quotients of an n-dimensional vector space. Let a rigid analytic space $\mathrm{Gr}(d,n)^{an}$ be the analytification [21, Ex. 4.3.3] of $\mathrm{Gr}(d,n)$.

Theorem 2.27 (Rapoport-Zink [45, Ch. 5]).   *There is an étale period morphism*

$$\pi_{RZ} : \mathcal{M}_{\mathbb{X},\eta} \longrightarrow \mathrm{Gr}(d,n)^{an}$$

of rigid analytic spaces over L.

Sketch of proof.   We describe the period morphism on L-valued points. Note that there is an equality $\mathcal{M}_{\mathbb{X},\eta}(L) = \mathcal{M}_{\mathbb{X}}(\mathcal{O}_L)$. The definition of $\mathcal{M}_{\mathbb{X}}$ yields

$$\mathcal{M}_{\mathbb{X}}(\mathcal{O}_L) = \left\{ \begin{array}{c} p\text{-divisible group } X \text{ over } \mathcal{O}_L \text{ with a} \\ \text{quasi-isogeny } \rho \colon X \times_{\mathrm{Spec}(\mathcal{O}_L)} \mathrm{Spec}(\bar{\mathbb{F}}_p) \dashrightarrow \mathbb{X} \end{array} \right\} \Big/ \simeq.$$

Choose a point $x \in \mathcal{M}_{\mathbb{X},\eta}(L)$ with the corresponding pair (X,ρ) and, for short, put $X_p := X \times_{\mathrm{Spec}(\mathcal{O}_L)} \mathrm{Spec}(\bar{\mathbb{F}}_p)$.

Since $\mathcal{O}_L \to \bar{\mathbb{F}}_p$ is a PD-thickening, (a variant of) Theorem 2.19(i) tells us that we have a canonical isomorphism $\mathrm{Lie}(EX) \simeq M(X_p)$ of free $\mathcal{O}_L$-modules of rank n, where $M(X_p)$ is the Dieudonné crystal of X_p restricted to $\mathcal{O}_L$. The quasi-isogeny ρ defines an isomorphism $\rho \colon M(X_p)_L \xrightarrow{\sim} M(\mathbb{X})_L$. The choice of X gives a quotient $\mathrm{Lie}(EX) \longrightarrow \mathrm{Lie}(X)$ of rank d. Altogether, this defines a d-dimensional quotient $\mathrm{Lie}(X)_L$ of the n-dimensional L-vector space $M(\mathbb{X})_L$. Now $\pi_{RZ}(x)$ is the corresponding point in $\mathrm{Gr}(d,n)^{an}(L) = \mathrm{Gr}(d,n)(L)$.   □

The period morphism π_{RZ} was already known in some cases, in particular in the Lubin-Tate case, before the work of Rapoport and Zink. Here are some examples:

Example 2.28 (Dwork).   Deformation space of an ordinary elliptic curve (see, e.g., [1, Sect. II.6.3.1]).

Example 2.29 (Lubin-Tate case). If $\mathbb{X}$ is the p-divisible group associated with a Lubin-Tate formal group of height n, then π_{RZ} coincides with the Gross-Hopkins period morphism [22]

$$\pi_{GH} : \mathcal{M}_{LT} = \bigsqcup_{n \in \mathbb{Z}} \mathbb{D}_L^{n-1} \longrightarrow (\mathbb{P}_L^{n-1})^{an},$$

where $\mathbb{D}_L^{n-1}$ is the open $(n-1)$-dimensional rigid analytic ball. The restriction of π_{GH} to each ball gives a surjective infinitely sheeted étale covering map.

In general, the nonempty fibers of π_{RZ} are isomorphic to $\boldsymbol{GL}_n(\mathbb{Q}_p)/\boldsymbol{GL}_n(\mathbb{Z}_p)$ (see [45, Prop. 5.37]).

2.4.6   Level Structures and a Conjecture of Kottwitz

Let $\mathbb{X}$ be a p-divisible group of dimension d and height n over $\bar{\mathbb{F}}_p$. We now introduce level structures. The procedure is literally the same as that for Lubin-Tate spaces sketched above and therefore we refer to Section 2.3.2 for details in order to omit unnecessary repetitions. The outcome is again a tower of finite étale coverings

$$\mathcal{M}_{\mathbb{X},m} \longrightarrow \mathcal{M}_{\mathbb{X},\eta}, \qquad m \geq 1,$$

and there is an action of $\boldsymbol{GL}_n(\mathbb{Q}_p)$ on the corresponding pro-object $\mathcal{M}_{\mathbb{X},\infty}$ in the category of rigid analytic spaces. The period morphism $\pi_{RZ} : \mathcal{M}_{\mathbb{X},\eta} \to \mathrm{Gr}(d,n)^{an}$ defines a map $\mathcal{M}_{\mathbb{X},\infty} \to \mathrm{Gr}(d,n)^{an}$, which we denote also by π_{RZ}.

The "philosophy" is that $\mathcal{M}_{\mathbb{X},\infty}$, if given a reasonable geometric sense, is a $\boldsymbol{GL}_n(\mathbb{Z}_p)$-torsor over $\mathcal{M}_{\mathbb{X},\eta}$ so that $\mathcal{M}_{\mathbb{X},\infty}$ is a $\boldsymbol{GL}_n(\mathbb{Q}_p)$-torsor over the (open) image of π_{RZ}. The appropriate geometric structure to put on $\mathcal{M}_{\mathbb{X},\infty}$ for this to work is that of a preperfectoid adic space [50, Thm. 6.3.4]. In this setting, one can also make sense of the "isomorphism" between the Lubin-Tate and the Drinfeld tower [14], [19] at the "infinite level." This is particularly noteworthy as there is no such isomorphism if the two towers are merely considered as pro-objects of rigid analytic spaces. The result of Scholze and Weinstein (cf. earlier work of Faltings and Fargues) takes the form of a duality isomorphism between two "dual" Rapoport-Zink spaces at the infinite level, which are both constructed like our $\mathcal{M}_{\mathbb{X},\infty}$ above. We refrain from giving details and refer to [50, Thm. 7.2.3] instead.

Let us now come back to the construction of L-parameters. For this, we do not need any perfectoid spaces and consider $\mathcal{M}_{\mathbb{X},\infty}$ just as a tower of rigid analytic spaces. As in the Lubin-Tate case, we have three actions on $\mathcal{M}_{\mathbb{X},\infty}$:

(i)   an action of $\boldsymbol{GL}_n(\mathbb{Q}_p)$;

(ii) an action of the group $\mathrm{QIsog}(\mathbb{X})$ of quasi-isogenies from $\mathbb{X}$ to itself;

(iii) an action of the Weil group $W_{\mathbb{Q}_p}$, which arises from a Weil descent datum $\mathcal{M}_{\mathbb{X},m} \xrightarrow{\sim} \mathcal{M}_{\mathbb{X},m} \times_{L,\sigma} L$ induced by the Frobenius morphism $F: \mathbb{X} \to \mathbb{X} \times_{\bar{\mathbb{F}}_p,\sigma} \bar{\mathbb{F}}_p$.

One can show that the group $\mathrm{QIsog}(\mathbb{X})$ can be identified with the $\mathbb{Q}_p$-points of a reductive group J_b over $\mathbb{Q}_p$. For instance, if $\mathbb{X}$ corresponds to the simple isocrystal with slope $\lambda \in [0,1]$, then J_b is the group of units $D^\times$ in the central division algebra D over $\mathbb{Q}_p$ with invariant $\lambda \in \mathbb{Q}/\mathbb{Z}$. In more explicit terms, any $\bar{\mathbb{F}}_p$-isocrystal can be written as $(L^n, b \circ \sigma)$ for some $b \in \boldsymbol{GL}_n(L)$. A quasi-isogeny is just an element of the σ-centralizer

$$\{g \in \boldsymbol{GL}_n(L) \mid g^{-1}b\sigma(g) = b\}.$$

In generalization of Section 2.3.2, for any prime l different from p, we get an action of the product $\boldsymbol{GL}_n(\mathbb{Q}_p) \times J_b(\mathbb{Q}_p) \times W_{\mathbb{Q}_p}$ on

$$H_c^*(\mathcal{M}_{\mathbb{X},\infty}) = \varinjlim_m H_{c,\acute{e}t}^*(\mathcal{M}_{\mathbb{X},m,\mathbb{C}_p}, \bar{\mathbb{Q}}_l).$$

The following conjecture says that supercuspidal representations can be realized in this cohomology group.

Conjecture 2.30 (Kottwitz). *Assume $\mathbb{X}$ isoclinic, that is, has just one slope, which is $\frac{d}{n}$. Let π be a supercuspidal representation of $GL_n(\mathbb{Q}_p)$. Then, we have an identity*

$$H_c^*(\mathcal{M}_{\mathbb{X},\infty})[\pi] = (-1)^{d(n-d)} JL(\pi) \otimes \bigwedge^d LLC(\pi)$$

in the Grothendieck group of representations of the group $J_b(\mathbb{Q}_p) \times W_{\mathbb{Q}_p}$ whose restriction to $J_b(\mathbb{Q}_p)$ is smooth admissible and whose restriction to $W_{\mathbb{Q}_p}$ is continuous.

This conjecture is now essentially a theorem of Fargues [18]. To understand its meaning, first note that the assumption on $\mathbb{X}$ implies that $J_b(\mathbb{Q}_p)$ is the group of units $D^\times$ in the central simple $\mathbb{Q}_p$-algebra D with invariant d/n and of dimension n^2, which is an inner form of $\boldsymbol{GL}_n$. The Jacquet-Langlands correspondence [9], [46] asserts a bijection

$$JL : \{\text{irreducible discrete series representations of } \boldsymbol{GL}_n(\mathbb{Q}_p)\}/_{\simeq} \xleftrightarrow{1:1}$$
$$\{\text{irreducible discrete series representations of } D^\times\}/_{\simeq}.$$

Also, recall that the local Langlands correspondence asserts a bijection

$$LLC : \{\text{supercuspidal representations of } \boldsymbol{GL}_n(\mathbb{Q}_p)\}/_{\simeq} \xleftrightarrow{1:1}$$
$$\{\text{irreducible } n\text{-dimensional representations of } W(\mathbb{Q}_p)\}/_{\simeq}.$$

It should be mentioned that, in general, $JL(\pi)$ is a smooth infinite-dimensional representation of $J_b(\mathbb{Q}_p)$, whereas $LLC(\pi)$ is always finite-dimensional. Fargues proves the above conjecture only when $(d,n) = 1$, in which case the first term $JL(\pi)$ is actually finite-dimensional.

One expects that π appears only in cohomology of degree $d(n-d)$.

2.4.7 Local Shimura Varieties

Comparing the definition of Rapoport-Zink spaces with that of Shimura varieties, it is clear that they are rather an analogue of PEL-type Shimura varieties than of general ones. Rapoport-Zink spaces and PEL-type Shimura varieties are constructed from moduli problems for p-divisible groups and abelian varieties, respectively, decorated with additional structures (polarization, endomorphism, level). Such a moduli problem is not available for general Shimura varieties and their definition resorts to the rather abstract notion of a Shimura datum. To adjust this imbalance, Rapoport and Viehmann [44] have suggested the existence of so-called local Shimura varieties generalizing Rapoport-Zink spaces. Like (global) Shimura varieties they should be determined by a purely group-theoretic datum without any underlying moduli problem.

Let G be a reductive group over $\mathbb{Q}_p$. A σ-conjugate of $b \in G(L)$ is any element of the form $g^{-1}b\sigma(g)$, where $g \in G(L)$. This gives an equivalence relation on $G(L)$ and the set of all such σ-conjugacy classes is commonly denoted by $B(G)$ in the literature. The set $B(G)$ is related to generalizations of the $\bar{\mathbb{F}}_p$-isocrystals introduced above. In fact, there is a straightforward correspondence between $B(\boldsymbol{GL}_n)$ and n-dimensional $\bar{\mathbb{F}}_p$-isocrystals as follows:

$$[b] \in B(\boldsymbol{GL}_n) \xleftrightarrow{1:1} \bar{\mathbb{F}}_p\text{-isocrystal } (L^n, b \circ \sigma).$$

The set $B(G)$ appears naturally in the reduction theory of Shimura varieties (or rather of their integral models) (see [43, Sect. 7]).

For each element $b \in G(L)$, define an algebraic group J_b over $\mathbb{Q}_p$ functorially by demanding

$$J_b(R) = \{g \in G(R \otimes_{\mathbb{Q}_p} L) \mid g^{-1}b\sigma(g) = b\}.$$

This generalizes the group of the same name in Section 2.4.6. If b, b' are σ-conjugate, then J_b and $J_{b'}$ are isomorphic.

In addition, the presence of a $G(\bar{\mathbb{Q}}_p)$-conjugacy class $\bar{\mu}$ of a cocharacter $\mu \colon \mathbb{G}_m \to G_{\bar{\mathbb{Q}}_p}$ singles out a subset $B(G, \bar{\mu})$ of μ-admissible conjugacy classes in $B(G)$ (see [43, Prop. 4.4]). We cannot discuss the definition of μ-admissibility here and just refer to the original definition in [36, Sect. 6.2].

Definition 2.31 ([44, Def. 5.1]). A *local Shimura datum* is a collection $(G, b, \bar{\mu})$ as above such that the following conditions are satisfied:

(i) μ is a minuscule cocharacter, that is, the induced action on $\mathrm{Lie}(G)$ has only weights in $\{-1, 0, 1\}$;

(ii) $[b] \in B(G, \{\mu\})$ (μ-admissibility).

Note that a (global) Shimura datum contains some homomorphism $h \colon S^1 \to G_{\mathbb{R}}$ with $S^1 = \mathrm{Res}_{\mathbb{C}/\mathbb{R}} \mathbb{G}_m$ the Deligne torus. Here, we just have a cocharacter $\mu \colon \mathbb{G}_m \to G_{\bar{\mathbb{Q}}_p}$.

In the sequel, the field of definition of the conjugacy class $\bar{\mu}$ is denoted by E and called the *reflex field* of the datum. Rapoport and Viehmann expressed the hope that from a datum as above a tower of rigid-analytic spaces can be constructed, much in the style of the tower of algebraic varieties constructed out of a Shimura datum. We recollect some of their expectations here:

Conjecture 2.32 ([44, Sect. 5.1]). *Let $(G, [b], \bar{\mu})$ be a local Shimura datum. For each compact open subgroup K of $G(\mathbb{Q}_p)$, there exists a rigid analytic space $\mathcal{M}_K = \mathcal{M}_{(G,b,\bar{\mu}),K}$ over $\breve{E} := E \cdot L$ and for each inclusion $K' \subset K$ of compact open subgroups in $G(\mathbb{Q}_p)$, there is a morphism $\mathcal{M}_{K'} \to \mathcal{M}_K$ such that the following hold true:*

- *(i) the group $J_b(\mathbb{Q}_p)$ acts on each $\mathcal{M}_K$ compatibly with the morphisms $\mathcal{M}_{K'} \to \mathcal{M}_K$;*
- *(ii) the group $G(\mathbb{Q}_p)$ acts on the pro-object $\mathcal{M}_\infty = \varprojlim_K \mathcal{M}_K$;*
- *(iii) there is a Weil descent datum $\mathcal{M}_\infty \xrightarrow{\sim} \mathcal{M}_\infty \times_{\breve{E},\sigma} \breve{E}$, where σ is induced by the Frobenius on L;*
- *(iv) there is a system of compatible rigid analytic étale period morphisms $\pi_K : \mathcal{M}_K \to \mathcal{F}l_{\bar{\mu}}$, where $\mathcal{F}l_{\bar{\mu}}$ parametrizes conjugacy classes of parabolics determined by $\bar{\mu}$.*

Of course, the conjecture is tailored in a way that Rapoport-Zink spaces give examples of local Shimura varieties. The case we considered above corresponds to $G = \boldsymbol{GL}_n$ and $\bar{\mu}$ being the conjugacy class of

$$\mu : \mathbb{G}_m \longrightarrow G_{\bar{\mathbb{Q}}_p} \qquad t \longmapsto \begin{pmatrix} t & & & & & & \\ & \ddots & & & & & \\ & & t & & & & \\ & & & 1 & & & \\ & & & & \ddots & & \\ & & & & & 1 \end{pmatrix}$$

with d times t and $n - d$ times 1 on the main diagonal.

Conjecture 2.32 is now an unpublished theorem, by the work of Fargues, Kedlaya-Liu, and Caraiani-Scholze. Hence, for each local Shimura datum $(G, b, \bar{\mu})$ we have once again a pro-object $\mathcal{M}_{(G,b,\bar{\mu}),\infty} = \varprojlim_K \mathcal{M}_{(G,b,\bar{\mu}),K}$ whose compactly supported étale cohomology groups

$$H_c^*(\mathcal{M}_{(G,b,\bar{\mu}),\infty}) = \varinjlim_K H_{c,\acute{e}t}^*(\mathcal{M}_{(G,b,\bar{\mu}),K} \times_{\breve{E}} \mathbb{C}_p, \bar{\mathbb{Q}}_l)$$

carry an action of $G(\mathbb{Q}_p) \times J_b(\mathbb{Q}_p) \times W_E$ that is smooth in the first two factors and continuous in the last. Here, W_E is the Weil group associated with E whose action on étale cohomology comes from the Weil descent datum.

Conjecture 2.33 (Kottwitz, vaguely). *If π is a supercuspidal representation of $G(\mathbb{Q}_p)$, then the "Galois part" of $H_c^*(\mathcal{M}_{(G,b,\bar{\mu}),\infty})[\pi]$ is "given by"*

$$r_{\bar{\mu}} \circ \varphi_\pi |_{W_E},$$

where

(i) $\varphi_\pi \colon W_{\mathbb{Q}_p} \to {}^L G$ is the L-parameter associated with π by the local Langlands correspondence;

(ii) $r_{\bar{\mu}} \colon {}^L(G_E) = {}^L G^\circ \rtimes W_E \to \boldsymbol{GL}_N$ is the highest weight representation of ${}^L G^\circ$ for the conjugacy class of characters $\bar{\mu}^\vee$ of ${}^L G^\circ$ dual to the cocharacters of G in $\bar{\mu}$.

Unfortunately, knowing $r_{\bar{\mu}} \circ \varphi_\pi$ does not allow one to construct the L-parameter φ_π itself. For this reason, we have to use an approach that does not rely on a minuscule cocharacter μ.

2.4.8   Local Shimura Varieties: Further Developments

The picture described in the previous sections (except for the existence of general local Shimura varieties) was the situation around 2010. Since then several different developments have led to a complete reinterpretation:

- Rigid analytic spaces were replaced by Huber's adic spaces, among which one can find Scholze's perfectoid spaces, introduced later. Finally, these were replaced by so-called "diamonds."
- p-divisible groups were replaced by p-adic Hodge structures and then by "mixed-characteristic shtukas" or, alternatively, by vector bundles on the Fargues-Fontaine curve.
- Rapoport-Zink spaces were replaced by local Shimura varieties and then by moduli spaces of "mixed-characteristic shtukas" or, alternatively, by the stack of G-bundles on the Fargues-Fontaine curve.
- The local Langlands correspondence was replaced by Laurent Fargues's "geometric Langlands correspondence on the Fargues-Fontaine curve." This geometric correspondence is still to be rigorously stated. Implicitly, this uses moduli spaces of local shtukas over "$\mathrm{Spec}(\mathbb{Q}_p) \times_{\mathrm{Spec}(\mathbb{F}_1)} \mathrm{Spec}(\mathbb{Q}_p)$," and there is a relation to the work of V. Lafforgue.

References for all of this can be found in Weinstein's notes on Scholze's Berkeley course in 2014 [49] and Feng's informal notes on the AG in Oberwolfach in April 2016 [53].

2.5   SOME BASICS OF P-ADIC GEOMETRY

2.5.1   Various Topological Rings

Adic spaces are generalizations of both formal schemes (thus, in particular, of schemes) and rigid analytic spaces. Adic spaces are glued from affinoid adic spaces, which are associated with some topological rings.

By a ring, we always mean a commutative ring with a unit. By an I-adic ring R, we mean a ring R with the I-adic topology induced by an ideal $I \subset R$.

Definition 2.34.

(i) A subset $S \subset R$ of a topological ring R is called *bounded* if for any open neighborhood U of 0, there exists an open neighborhood V of 0 such that $V \cdot S \subset U$.

(ii) An element $x \in R$ of a topological ring R is called *power-bounded* if the set

$$\{x^n \mid n \in \mathbb{N}\} \subset R$$

is bounded. The set of all power-bounded elements of R is denoted by R°.

(iii) A *Huber ring*[4] is a topological ring R which admits an open subring $R_0 \subset R$, called a *ring of definition*, such that R_0 with the restricted topology from R is an I-adic ring for some finitely generated ideal $I \subset R_0$, called an *ideal of definition*.

We do not require our topological rings to be complete, but we can restrict ourselves to complete (separated) rings everywhere because the associated affinoid adic spaces of a ring and of its completion are the same. The fact that I is finitely generated is actually crucial to ensure good properties for the completion [28, Lem. 1.6].

Clearly, any finite subset of a topological ring R is bounded and any nilpotent element is power-bounded. If R admits a base of open neighborhood of 0 given by subgroups (with respect to the additive group structure), then in Definition 2.34(i), we may assume that U and V are subgroups and $V \cdot S$ denotes the subgroup generated by pairwise products of elements of V and S. This will be the case for what follows.

Note that the ring of definition and the ideal of definition are not defined uniquely for a Huber ring. For example, one can take any power of the ideal of definition.

Example 2.35.

(i) The field $\mathbb{Q}_p$ with the p-adic topology is a Huber ring with $\mathbb{Z}_p$ being the ring of definition. The ring $\mathbb{Z}_p$ is a Huber ring as well, being its own ring of definition.

(ii) Any I-adic ring R for some finitely generated ideal $I \subset R$ is a Huber ring (one can take $R_0 = R$ in this case). In particular, any discrete ring is a Huber ring, being a (0)-adic ring.

Let R be a Huber ring with a ring of definition R_0 and an ideal of definition $I \subset R_0$.

By definition, we have that the collection $\{I^n \mid n \in \mathbb{N}\}$ is a base of open neighborhoods of 0 in both R_0 and R. For any element $x \in R$, the multiplication by x is a continuous map from R to itself, whence there is $n \in \mathbb{N}$ such that

---

[4]Called an *f-adic ring* by Huber [28].

$I^n \cdot x \subset R_0$. It follows that $R = \bigcup_{n \in \mathbb{N}} I^{-n} R_0$. Also, one shows easily that a subset $S \subset R$ is bounded if and only if there is $n \in \mathbb{N}$ such that $I^n \cdot S \subset R_0$.

Proposition 2.36. *A subring $R_0' \subset R$ is a ring of definition if and only if it is open and bounded.*

Proof. One implication is evident. For the other implication, assume that $R_0' \subset R$ is an open bounded subring. After possibly replacing the ideal of definition by its power, we may assume that $I \subset R_0'$ and $I \cdot R_0' \subset R_0$. Then the ideal $J := I \cdot R_0'$ in the ring R_0' is open because $I \subset J$. On the other hand, we have the embedding

$$J^2 = I \cdot I \cdot R_0' \subset I \cdot R_0 = I.$$

This implies that $\{J^n \mid n \in \mathbb{N}\}$ is a base of open neighborhoods of 0 in R.

It remains to replace the ideal J by a finitely generated ideal. Namely, let $x_1, \ldots, x_r$ be a finite set of generators of I in R_0. Then the ideal I' of R_0' generated by $x_1, \ldots, x_r$ is open, because $I' = R_0' \cdot \{x_1, \ldots, x_r\}$ contains $I^2 = I \cdot \{x_1, \ldots, x_r\}$. Since I' is contained in J, we see that $\{(I')^n \mid n \in \mathbb{N}\}$ is also a base of open neighborhoods of 0 in R.   □

It follows from Proposition 2.36 that for any ring of definition $R_0' \subset R$, there is $n \in \mathbb{N}$ such that $R_0' \subset I^{-n} R_0$.

Lemma 2.37. *For any Huber ring R, the set of all power-bounded elements R° is the union of all rings of definition in R.*

Proof. Clearly, any ring of definition is contained in R°. Conversely, take an element $x \in R^\circ$. The subring $R_0[x] \subset R$ is open as it contains R_0. Since x is power-bounded, there is $m \in \mathbb{N}$ such that

$$I^m \cdot \{x^n \mid n \in \mathbb{N}\} \subset R_0.$$

Then $I^m \cdot R_0[x] \subset R_0$, whence $R_0[x]$ is bounded and we finish the proof by Proposition 2.36.   □

Note that the union in Lemma 2.37 is directed. Indeed, given a ring of definition R_0', consider the subring $R_0 \cdot R_0' \subset R$ generated by R_0 and R_0'. By Proposition 2.36, we have that $I^n \cdot R_0' \subset R_0$ for some $n \in \mathbb{N}$. Therefore, $I^n \cdot (R_0 \cdot R_0') \subset R_0$, whence the subring $R_0 \cdot R_0'$ is open bounded and again by Proposition 2.36, $R_0 \cdot R_0'$ is a ring of definition.

For example, for the Huber ring $\mathbb{Q}_p$, we have that $\mathbb{Q}_p^\circ = \mathbb{Z}_p$.

Definition 2.38. A Huber ring R is called *uniform* if the subring $R^\circ \subset R$ is bounded.

By Proposition 2.36, for a uniform Huber ring R, the subring R° is a ring of definition. Not all Huber rings are uniform; see Example 2.40 below.

Consider the following special class of Huber rings.

Definition 2.39. A Huber ring R is a *Tate ring* if it has a topologically nilpotent invertible element $\varpi \in R$, meaning that for any open neighborhood of zero $U \subset R$, there is $n \in \mathbb{N}$ such that $\varpi^i \in U$ for all $i \geqslant n$. We call such an element ϖ a *pseudo-uniformizer*.

For example, $\mathbb{Q}_p$ is a Tate ring while $\mathbb{Z}_p$ is not. Pseudo-uniformizers are not uniquely defined. In particular, one can replace ϖ by ϖ^n for any integer $n > 0$.

Now suppose that R is a Tate ring with a quasi-uniformizer $\varpi \in R$. Clearly, ϖ is power-bounded and the ring $R_0[\varpi]$ is a ring of definition. Moreover, there is an $n \in \mathbb{N}$ such that $\varpi^n \in R_0$. Thus, changing either R_0 or ϖ, we may always assume that $\varpi \in R_0$. Then $(\varpi) \subset R_0$ can be taken as an ideal of definition. Indeed, multiplication by ϖ^{-1} is a continuous map from R to itself, which implies that the ideal $(\varpi) \subset R_0$ is open. Using that ϖ is topologically nilpotent, one shows that the ϖ-adic topology on R_0 coincides with the given one.

The above discussion on Huber rings implies that there is an equality $R = R_0[\varpi^{-1}]$, that a subset $S \subset R$ is bounded if and only if there is an $n \in \mathbb{N}$ such that $\varpi^n \cdot S \subset R_0$, and that for any ring of definition R_0', there is an $n \in \mathbb{N}$ such that $R_0' \subset \varpi^{-n} R_0$.

Example 2.40. Consider the Tate ring $R = \mathbb{Q}_p[T]/(T^2) = \mathbb{Q}_p \oplus \mathbb{Q}_p \cdot T$ with the natural p-adic topology. Then for any $n \in \mathbb{N}$, we have that $\mathbb{Z}_p \oplus p^{-n}\mathbb{Z}_p \cdot T$ is a ring of definition. In addition, there is an equality

$$R^\circ = \mathbb{Z}_p \oplus \mathbb{Q}_p \cdot T = \bigcup_{n \geqslant 0} \mathbb{Z}_p \oplus p^{-n}\mathbb{Z}_p \cdot T.$$

In particular, the Tate ring $\mathbb{Q}_p[T]/(T^2)$ is not uniform.

This example motivates the following fact.

Lemma 2.41. *A separated uniform Tate ring R is reduced.*

Proof. Suppose that R is a separated uniform Tate ring and that there is a nonzero nilpotent element $x \in R$. Then all nilpotent elements $\varpi^i x, i \in \mathbb{Z}$, belong to R°. Since R° is bounded, the set $\{\varpi^i x \mid i \in \mathbb{Z}\}$ is bounded as well. Using that ϖ is invertible, we see that for any $n \in \mathbb{N}$, there is an embedding

$$\{\varpi^i x \mid i \in \mathbb{Z}\} \subset (\varpi)^n = \varpi^n \cdot R_0.$$

Since R is separated and (ϖ) is an ideal of definition of R_0, we obtain a contradiction. $\square$

Theorem 2.44 below is a sort of a converse to Lemma 2.41. Before stating it, we must introduce non-archimedean fields.

Definition 2.42. A *non-archimedean field* K is a non-discrete topological field whose topology is induced by a non-archimedean norm $\|\cdot\| \colon K \to \mathbb{R}_{\geqslant 0}$. The *ring of integers* of K is defined by the formula

$$\mathcal{O}_K := \{x \in K \mid \|x\| \leqslant 1\}.$$

The non-archimedean norm $\|\cdot\|$ in the above definition is not considered to be part of the data. It is only required that such a norm exists. However, the norm $\|\cdot\|$ is unique up to rescaling $t \mapsto t^a$ on $\mathbb{R}_{\geqslant 0}$ for any positive real a. In particular, the ring of integers is well-defined.

Let K be a non-archimedean field. The ring of integers $\mathcal{O}_K$ is a local ring with the maximal ideal

$$\mathfrak{m}_K = \{x \in K \mid \|x\| < 1\}.$$

Since K is not discrete by definition, we have that $\mathfrak{m}_K \neq 0$.

We claim that the topological field K is a Tate ring. Indeed, one can take for the ring of definition the ring of integers $\mathcal{O}_K$ and for the ideal of definition the principal ideal

$$I = (\varpi) = \{x \in R \mid \|x\| \leqslant \|\varpi\|\},$$

where $\varpi \in \mathfrak{m}_K$ is any nonzero element, exactly as was done for $K = \mathbb{Q}_p$. Furthermore, ϖ is a pseudo-uniformizer in K. Note that the $\|\cdot\|$-topology on $\mathcal{O}_K$ coincides with the I-adic topology, but not with the $\mathfrak{m}_K$-adic topology, in general. For example, it may be that $\mathfrak{m}_K^2 = \mathfrak{m}_K$, which is essential for Faltings's "almost mathematics" [13].

Clearly, a subset $S \subset K$ is bounded if and only if the subset $\|S\| \subset \mathbb{R}_{\geqslant 0}$ is bounded (above). Also, an additive subgroup of K is open if and only if it contains $\{x \in K \mid \|x\| < t\}$ for some $t \in \mathbb{R}_{> 0}$.

It is shown easily that for a non-archimedean field K, we have $K^\circ = \mathcal{O}_K$. In particular K is uniform.

Let R be a topological K-algebra. Then R is a Huber ring if and only if R is a Tate ring, with any pseudo-uniformizer ϖ in K also a pseudo-uniformizer in R. In this case, one can find a ring of definition of R which is an $\mathcal{O}_K$-subalgebra.

Example 2.43.

(i) Consider the algebra of polynomials $K[T_1, \ldots, T_n]$ with the natural ϖ-adic topology, that is, with a base of open neighborhoods of 0 given by $\pi^i \mathcal{O}_K[T_1, \ldots, T_n]$, $i \in \mathbb{Z}$. Then the topological K-algebra $K[T_1, \ldots, T_n]$ is a Tate ring with a ring of definition $\mathcal{O}_K[T_1, \ldots, T_n]$.

(ii) Define the topological K-algebra $K\langle T_1, \ldots, T_n \rangle$ as the completion of $K[T_1, \ldots, T_n]$ from (i). Explicitly, there is an equality

$$K\langle T_1, \ldots, T_n \rangle = \left\{ \sum_{l \in \mathbb{N}^n} a_l T^l \mid a_l \in K \text{ and } a_l \to 0 \text{ as } |l| \to +\infty \right\},$$

where we use multi-index notation, that is, for any $l = (l_1, \ldots, l_n) \in \mathbb{N}^n$, we put $T^l := T_1^{l_1} \ldots T_n^{l_n}$ and $|l| := l_1 + \ldots + l_n$. Then $K\langle T_1, \ldots, T_n \rangle$ is a Tate ring with a ring of definition $\mathcal{O}_K\langle T_1, \ldots, T_n \rangle$. One can show that the equality

$$K\langle T_1, \ldots, T_n \rangle^\circ = \mathcal{O}_K\langle T_1, \ldots, T_n \rangle$$

holds, whence $K\langle T_1, \ldots, T_n \rangle$ is uniform.

The following result is proved in [4, Prop. 6.2.3/1, Thm. 6.2.4/1].

Theorem 2.44. *Let K be a non-archimedean field. Let R be a complete reduced topological K-algebra which is topologically finitely generated over K, that is, there is an isomorphism of topological K-algebras $R \simeq K\langle T_1, \ldots, T_n \rangle / \mathfrak{a}$ for some radical ideal $\mathfrak{a} \subset R\langle T_1, \ldots, T_n \rangle$. Then R is uniform.*

Note that $\mathfrak{a}$ is automatically closed in $K\langle T_1, \ldots, T_n \rangle$ and finitely generated because the ring $K\langle T_1, \ldots, T_n \rangle$ is Noetherian. The quotient $K\langle T_1, \ldots, T_n \rangle / \mathfrak{a}$ is a Tate ring. One can take as a ring of definition the image of $\mathcal{O}_K\langle T_1, \ldots, T_n \rangle$ by Banach's open mapping theorem.

2.5.2 Continuous Valuations

Let Γ be a *totally ordered abelian group* (with the group law written multiplicatively). This means that we have a total order on Γ which is translation-invariant, that is, for all $\gamma, \gamma', \gamma'' \in \Gamma$ with $\gamma' \leqslant \gamma''$, we have that $\gamma\gamma' \leqslant \gamma\gamma''$. Recall the following notion.

Definition 2.45. A *valuation* on a ring R with values in $\Gamma \cup \{0\}$ is a map

$$|\cdot| : R \longrightarrow \Gamma \cup \{0\},$$

such that $|0| = 0$, $|1| = 1$, and for all $x, y \in R$, we have

$$|xy| = |x||y|, \qquad |x + y| \leqslant \max\{|x|, |y|\}.$$

By definition, we put $0 \cdot \gamma = 0$ and $0 < \gamma$ for any $\gamma \in \Gamma$. The *value group of* $|\cdot|$ is the subgroup $\text{Im}(|\cdot|) \smallsetminus \{0\}$ of Γ. The *kernel of* $|\cdot|$ is the subset

$$\text{Ker}(|\cdot|) := \{x \in R \mid |x| = 0\} \subset R.$$

Two valuations $|\cdot|_1$ and $|\cdot|_2$ are *equivalent* if there is an order preserving isomorphism between their value groups that commutes with $|\cdot|_1$ and $|\cdot|_2$.

Note that sometimes one might call $|\cdot|$ as in Definition 2.45 a (semi-)norm rather than a valuation.

Clearly, $\mathrm{Ker}(|\cdot|)$ is a prime ideal of R and $|\cdot|$ factors uniquely through a valuation

$$|\cdot|' : R/\mathrm{Ker}(|\cdot|) \longrightarrow \Gamma \cup \{0\}.$$

A valuation $|\cdot|$ defines a $|\cdot|$-*topology* on R with the base of open neighborhoods of 0 given by the subsets $\{x \in R \mid |x| < \gamma\} \subset R$, where $\gamma \in \Gamma$. An example is given by a non-archimedean topological field K with the norm $\|\cdot\|$.

Definition 2.46. A valuation $|\cdot|: R \to \Gamma \cup \{0\}$ on a topological ring R is *continuous* if the identity map from R with the given topology to R with the $|\cdot|$-topology is continuous.

Explicitly, we have that $|\cdot|$ is continuous if for any $\gamma \in \Gamma$, the subset

$$\{x \in R \mid |x| < \gamma\} \subset R$$

is open.

Example 2.47. The set of all continuous valuations $|\cdot|: R \to \Gamma \cup \{0\}$ with the trivial group $\Gamma = \{1\}$ is naturally in bijection with the set of all open prime ideals of R. The bijection sends such a continuous valuation to its kernel.

We will use the following simple fact.

Lemma 2.48. *Let $\gamma_0 \in \Gamma$ be such that $\gamma_0 < 1$ and for any $\gamma \in \Gamma$, there is $n \in \mathbb{N}$ such that $\gamma_0^n < \gamma$. Choose a real number t_0 with $0 < t_0 < 1$. Then there is a unique order preserving group homomorphism $\lambda: \Gamma \to \mathbb{R}_{>0}$ such that $\lambda(\gamma_0) = t_0$.*

Proof. The condition on γ_0 implies that for any element $\gamma \in \Gamma$, there is an integer $n \in \mathbb{Z}$ such that $\gamma_0^n \leqslant \gamma < \gamma_0^{n+1}$. Denote this integer by $[\gamma]$. Then λ sends $\gamma \in \Gamma$ to t_0 raised to the exponent $\lim_{i \to \infty} \frac{1}{i}[\gamma^i]$. $\qquad\square$

Corollary 2.49. *Let R be a topological ring and let $|\cdot|: R \to \Gamma \cup \{0\}$ be a continuous valuation. Suppose that there is a nonzero topologically nilpotent element $x_0 \in R$ such that $|x_0| \neq 0$. Then there is a unique up to rescaling continuous valuation $\|\cdot\|: R \to \mathbb{R}_{\geqslant 0}$ such that $|\cdot|$ is continuous with respect to the $\|\cdot\|$-topology and $\|\cdot\|$ factors as the composition of $|\cdot|$ with an order preserving homomorphism $\Gamma \to \mathbb{R}_{>0}$. Moreover, the $|\cdot|$-topology coincides with the $\|\cdot\|$-topology, and $\mathrm{Ker}(|\cdot|) = \mathrm{Ker}(\|\cdot\|)$.*

Proof. Applying Lemma 2.48 with $\gamma_0 = |x_0|$ and with any t_0, $0 < t_0 < 1$, we obtain a valuation $\|\cdot\| = \lambda \circ |\cdot|$ that satisfies all properties from the corollary.

To prove uniqueness use again Lemma 2.48. $\qquad\square$

Proposition 2.50. *Let R be a Tate ring and let $|\cdot|: R \to \Gamma \cup \{0\}$ be a continuous valuation. Then there exists a unique factorization of $|\cdot|$ as a composition*

$$R \longrightarrow K(|\cdot|) \overset{|\cdot|'}{\longrightarrow} \Gamma \cup \{0\},$$

where $K(|\cdot|)$ is a complete non-archimedean field such that the fraction field of the image of the homomorphism $R \to K(|\cdot|)$ is dense and $|\cdot|'$ is a continuous valuation on the non-archimedean field $K(|\cdot|)$.

Proof. Since a pseudo-uniformizer $\varpi \in R$ is invertible, it follows that $|\varpi| \neq 0$. Applying Corollary 2.49 to the continuous valuation $|\cdot|$ and ϖ, we obtain a valuation $\| \cdot \| \colon R \to \mathbb{R}_{\geqslant 0}$ with all properties from the corollary.

Since $\mathrm{Ker}(|\cdot|) = \mathrm{Ker}(\| \cdot \|)$, one factors uniquely the valuations $|\cdot|$ and $\| \cdot \|$ through valuations $|\cdot|'$ and $\| \cdot \|$ on the integral domain $R/\mathrm{Ker}(|\cdot|)$ and then on its field of fractions F. Finally, let $K(|\cdot|)$ be the completion of F with respect to the valuation $|\cdot|'$ (or, equivalently, to the valuation $\| \cdot \|$). Since the image of ϖ in $K(|\cdot|)$ is a nonzero topologically nilpotent element, the topological field $K(|\cdot|)$ is non-discrete, whence $K(|\cdot|)$ is a non-archimedean field with the norm $\| \cdot \|$.

Uniqueness follows from Corollary 2.49 applied to $K(|\cdot|)$. $\qquad\square$

According to Proposition 2.50, in order to describe all continuous valuations $|\cdot|$ on a Tate ring that factor through a given non-archimedean field $K(|\cdot|)$, one needs to describe all continuous valuations on $K(|\cdot|)$.

Recall that a subring K^+ of a field K is called a *valuation subring* if for any nonzero $x \in K$, either x or x^{-1} is contained in K^+. There is a natural bijection between the set of equivalence classes of valuations on K and the set of valuation subrings. Namely, a valuation $|\cdot|$ corresponds to the valuation subring $\{x \in K \mid |x| \leqslant 1\}$ and a valuation subring $K^+ \subset K$ corresponds to a valuation on K with the value group $K^*/(K^+)^*$, where the order is defined by $\bar{x} \leqslant \bar{y}$ if there is $z \in K^+$ such that $x = yz$, where $x, y \in K^*$ and $\bar{x}, \bar{y} \in K^*/(K^+)^*$ are the classes of x, y.

Lemma 2.51. *Let K be a non-archimedean field. Then the following is true:*

(i) *There is a bijection between open bounded valuation subrings of K and valuation subrings of the residue field $\kappa_K := \mathcal{O}_K/\mathfrak{m}_K$ of K. The bijection sends an open bounded valuation subring $K^+ \subset K$ to $K^+/\mathfrak{m}_K \subset \kappa_K$.*

(ii) *There is a bijection between continuous valuations of K up to equivalence and open bounded valuation subrings of K. The bijection sends a valuation $|\cdot|$ to the subring*

$$K^+ := \{x \in K \mid |x| \leqslant 1\}.$$

Proof. (*i*) By Proposition 2.36 and Lemma 2.37, any open bounded subring of K is contained in $K^\circ = \mathcal{O}_K$. Using that the ring $\mathcal{O}_K$ is local, one checks easily that if K^+ is an open bounded valuation subring of K, then $K^+ \supset \mathfrak{m}_K$.

(ii) Let $|\cdot|$ be a continuous valuation on K. By Corollary 2.49, the norm $\|\cdot\|$ on K (up to rescaling) factors through the valuation $|\cdot|$. This implies that K^+ is bounded and open.

Conversely, let K^+ be an open bounded valuation subring of K. By (i), we have that $\mathfrak{m}_K \subset K^+ \subset \mathcal{O}_K$. Thus the norm $\|\cdot\|\colon K^* \to K^*/\mathcal{O}_K^* \subset \mathbb{R}_{>0}$ factors through the valuation $|\cdot|\colon K^* \to K^*/(K^+)^*$, which implies that $|\cdot|$ is continuous. $\square$

Example 2.52.   Consider the Tate ring $\mathbb{Q}_p\langle T\rangle$ and its continuous valuation given by the Gauss norm

$$|\cdot|_{\mathrm{Gau\ss}} : \mathbb{Q}_p\langle T\rangle \longrightarrow p^{\mathbb{Z}} \cup \{0\},$$

$$\sum_{n\geq 0} a_n T^n \longmapsto \max_{n\geq 0} |a_n|_p,$$

where we normalize by $|p|_p := p^{-1}$. The name of the norm is derived from the fact that one uses the Gauss lemma in order to check that $|\cdot|_{\mathrm{Gau\ss}}$ is multiplicative.

Proposition 2.50 associates to the continuous valuation $|\cdot|_{\mathrm{Gau\ss}}$ the complete non-archimedean field $K(|\cdot|_{\mathrm{Gau\ss}})$, which is the field of fractions $\mathbb{Q}_p\{T\} := \mathbb{Q}_p\langle T\rangle[T^{-1}]$ of $\mathbb{Q}_p\langle T\rangle$ with the p-adic valuation given by the Gauss norm and with the residue field $\kappa_{K(|\cdot|_{\mathrm{Gau\ss}})} = \mathbb{F}_p(T)$. In more geometric terms, $\mathbb{Q}_p\{T\}$ is the completion of the local ring of the closed fiber $\mathbb{P}^1_{\mathbb{F}_p}$ of the scheme $\mathbb{P}^1_{\mathbb{Z}_p}$ over $\mathrm{Spec}(\mathbb{Z}_p)$.

According to Lemma 2.51, continuous valuations on $K(|\cdot|_{\mathrm{Gau\ss}})$ are in bijection with valuation subrings of the field $\mathbb{F}_p(T)$, that is, with the set of all schematic points on the projective line $\mathbb{P}^1_{\mathbb{F}_p}$ over $\mathbb{F}_p$. Under this bijection, the generic point of $\mathbb{P}^1_{\mathbb{F}_p}$ corresponds to the Gauss norm itself. Further, the closed point $\infty \in \mathbb{P}^1_{\mathbb{F}_p}$ corresponds to the continuous valuation $|\cdot|_\infty$ on $\mathbb{Q}_p\langle T\rangle$ with the value group $p^{\mathbb{Z}} \times \gamma^{\mathbb{Z}} \cup \{0\}$, where $\gamma > 1$ and for all $n \geq 1$, we put $\gamma^n < p$ (this is a rank two valuation). In other words, we consider a lexicographical order on $p^{\mathbb{Z}} \times \gamma^{\mathbb{Z}} \simeq \mathbb{Z} \times \mathbb{Z}$. The valuation is defined by the formula

$$|\sum_{n\geq 0} a_n T^n|_\infty := \max_{n\geq 0}(|a_n|\gamma^n).$$

In particular, we have that $|T|_\infty = \gamma > 1$. Notice that the continuous valuation $|\cdot| = |\cdot|_\infty$ on the non-archimedean field $\mathbb{Q}_p\langle T\rangle$ satisfies the following property: there is a power-bounded element $x \in \mathbb{Q}_p\langle T\rangle^\circ = \mathbb{Z}_p\langle T\rangle$ such that $|x| > 1$ (for example, one can take $x = T$). This is only possible when $|x| \in \Gamma$ is infinitesimally close to 1, that is, when $|x|$ belongs to the kernel of the homomorphism $\Gamma \to \mathbb{R}_{>0}$ as Corollary 2.49.

2.5.3 Adic Spaces and Perfectoids

Let R be a Huber ring, and let R^+ be an open subring of R° which is integrally closed in R.

Definition 2.53. We let
$$\mathrm{Spa}(R, R^+)$$
be the set of all equivalence classes of continuous valuations $|\cdot|$ on R such that $|x| \leqslant 1$ for all $x \in R^+$. We endow $\mathrm{Spa}(R, R^+)$ with the topology generated by the subsets
$$U_{f,g} := \{|\cdot| \, | \, |f| \leqslant |g| \neq 0\}$$
with f, g varying in R.

Note that the topological space $X := \mathrm{Spa}(R, R^+)$ comes equipped with two presheaves of topological rings $\mathcal{O}_X \supset \mathcal{O}_X^+$ [29].

Definition 2.54. If $\mathcal{O}_X$ (and hence also $\mathcal{O}_X^+$) is a sheaf, then we say that $X = \mathrm{Spa}(R, R^+)$ is an *affinoid adic space*.

One defines *adic spaces* by a gluing procedure out of affinoid adic spaces.

Example 2.55.

(i) A possible choice for R^+ is the subring R° (one checks easily that R° is integrally closed in R). We will write $\mathrm{Spa}(R)$ for $\mathrm{Spa}(R, R^\circ)$.
(ii) If K is a non-archimedean field, then $\mathrm{Spa}(K)$ is one point by Lemma 2.51.

Now fix a prime number p.

Definition 2.56. A *perfectoid ring* is a complete Tate ring R such that R is bounded and there is a pseudo-uniformizer $\varpi \in R$ such that ϖ^p divides p in R° (that is, $p = r\varpi^p$ for some $r \in R^\circ$) and the Frobenius map
$$\Phi : R^\circ/\varpi \longrightarrow R^\circ/\varpi^p, \qquad x \longmapsto x^p,$$
is an isomorphism.

Using the fact that R° is integrally closed in R, one shows that Φ is injective (provided that ϖ^p divides p in R°). Thus the condition that Φ is an isomorphism is equivalent to surjectivity of the Frobenius map on the ring R°/ϖ^p of characteristic p.

Example 2.57.

(i) The non-archimedean field $\mathbb{C}_p$ is perfectoid. Indeed, we have $\mathbb{C}_p^\circ = \mathcal{O}_{\mathbb{C}_p}$ and we can take $\varpi = p^{1/p}$.

(ii) The $\mathbb{C}_p$-algebra $\mathbb{C}_p\langle T\rangle$ is not perfectoid, because the Frobenius map is not surjective on the ring $(\mathcal{O}_K/p)[T]$. Let $\mathcal{O}_{\mathbb{C}_p}\langle T^{1/p^\infty}\rangle$ be the p-adic completion of the ring

$$\mathcal{O}_{\mathbb{C}_p}[T^{1/p^\infty}] := \bigcup_{r \geqslant 1} \mathcal{O}_{\mathbb{C}_p}[T^{1/p^r}].$$

Then the ring $\mathbb{C}_p\langle T^{1/p^\infty}\rangle := \mathcal{O}_{\mathbb{C}_p}\langle T^{1/p^\infty}\rangle[1/p]$ is perfectoid.

We remark that if R is perfectoid, then $X = \mathrm{Spa}(R, R^+)$ is an affinoid adic space, that is, the presheaf $\mathcal{O}_X$ is a sheaf [48, Thm. 6.3]. These are the building blocks of perfectoid spaces. Perfectoid spaces form a full subcategory of adic spaces.

2.6   APPROACH TO LLC VIA FARGUES-FONTAINE CURVE

2.6.1   Fargues-Fontaine Curve

Let us fix a complete algebraically closed non-archimedean field C of characteristic p. For example, one can take the completion $\widehat{\overline{\mathbb{F}_p((t))}}$ of the algebraic closure $\overline{\mathbb{F}_p((t))}$ of $\mathbb{F}_p((t))$.

Definition 2.58.   Let A_{inf} be the topological ring of Witt vectors $W(\mathcal{O}_C)$ endowed with the $(p, [\varpi])$-adic topology, where $[\varpi]$ denotes the Teichmüller representative of a pseudo-uniformizer $\varpi \in C$.

The ring A_{inf} is similar to a complete two-dimensional regular local ring, but it is not Noetherian (because $\mathcal{O}_C$ is not Noetherian due to the presence of all the p-th roots of ϖ). Note that the topology on A_{inf} is well-defined, that is, does not depend on the choice of a pseudo-uniformizer ϖ. Clearly, A_{inf} is a Huber ring, being a $(p, [\varpi])$-adic ring, and we have $A_{\mathrm{inf}}^\circ = A_{\mathrm{inf}}$.

The ring $W(\overline{\mathbb{F}}_p)$ is canonically embedded into A_{inf}. By definition, we have a Frobenuis endomorphism $\varphi \colon A_{\mathrm{inf}} \to A_{\mathrm{inf}}$, whose subring of invariants is $\mathbb{Z}_p = W(\mathbb{F}_p)$.

We define

$$Y_{FF} := \mathrm{Spa}(A_{\mathrm{inf}}) \smallsetminus \{p \cdot [\varpi] = 0\}.$$

It is an adic space over $L := W(\overline{\mathbb{F}}_p)[\frac{1}{p}]$ (but not an affinoid adic space) with a canonical action of the Frobenius φ induced by its action on A_{inf}. This action is proper and totally discontinuous.

Definition 2.59.   The *adic Fargues-Fontaine curve* is the quotient

$$X_{FF} := Y_{FF}/\varphi^{\mathbb{Z}}.$$

The curve X_{FF} is an adic space. It behaves like a smooth projective curve over an algebraically closed field. Indeed, Y_{FF} shares many features with a pointed open ball $\mathbb{D}^\circ$ over a non-archimedean field K and the curve X_{FF} bears some resemblance with the Tate curve obtained by quotienting out $\mathbb{D}^\circ$ by $\varpi^{\mathbb{Z}}$. In the adic setting, the pointed open ball over K is isomorphic to

$$\mathrm{Spa}(\mathcal{O}_K[[t]]) \smallsetminus \{t \cdot \varpi = 0\},$$

where $\mathcal{O}_K[[t]]$ is endowed with the (t, ϖ)-adic topology, and ϖ is a pseudo-uniformizer in K.

On the other hand, X_{FF} is not of finite type over any algebraically closed field, as we can see, for instance, by computing the global sections of the structure sheaf: there is an equality $H^0(X_{FF}, \mathcal{O}_{X_{FF}}) = \mathbb{Q}_p$.

There is a natural functor

$$\{\bar{\mathbb{F}}_p\text{-isocrystals}\} \longrightarrow \left\{ \begin{array}{c} \varphi\text{-equivariant} \\ \text{vector bundles on } Y_{FF} \end{array} \right\} \xrightarrow{\sim} \{\text{vector bundles on } X_{FF}\}.$$

The functor sends an isocrystal (W, φ_W) to the φ-equivariant vector bundle $(W \otimes_L \mathcal{O}_{Y_{FF}}, \varphi_W \otimes \varphi)$. We denote by $\mathcal{O}(\lambda)$ the vector bundle on X_{FF} associated to the simple isocrystal of slope λ. The following result is by Fargues and Fontaine [17, Théor. 8.2.10].

Theorem 2.60.

(i) *The functor above is essentially surjective and induces a bijection between the sets of isomorphism classes*

$$\{\bar{\mathbb{F}}_p\text{-isocrystals}\}/_{\simeq} \xleftrightarrow{1:1} \{\text{vector bundles on } X_{FF}\}/_{\simeq}.$$

(ii) *There is a canonical isomorphism $\pi_1^{\acute{e}t}(X_{FF}) \simeq G_{\mathbb{Q}_p} := \mathrm{Gal}(\overline{\mathbb{Q}}_p/\mathbb{Q}_p)$.*

Notice that the functor in Theorem 2.60(i) is not an equivalence of categories! There are no nonzero homomorphisms between isocrystals of different slopes, but they may arise for the associated vector bundles. Also, on isocrystals the canonical filtrations by slopes are canonically split, and this is not true in general on the vector bundles' side.

For every finite extension $\mathbb{Q}_p \subset K$ one can define an étale covering $X_{FF,K}$ of X_{FF}. Elaborating on Theorem 2.60(ii), we can prove that such curves are the only finite étale coverings of X_{FF}. In addition, the isomorphism on Galois groups translates Poincaré duality into local Tate duality [16, Cor. 3.8].

2.6.2 Fargues-Fontaine Curve and p-divisible Groups

Here is a construction of a field C as in the previous subsection. Start with $\mathbb{C}_p$ (or, more generally, with any complete algebraically closed extension of $\mathbb{Q}_p$). Following Fontaine, consider

$$\mathbb{C}_p^\flat := \varprojlim_{x \mapsto x^p} \mathbb{C}_p = \{(\ldots, a^{(1)}, a^{(0)}) \mid a^{(i)} \in \mathbb{C}_p, (a^{(i+1)})^p = a^{(i)}\}.$$

By construction, this is a multiplicative monoid. It turns out that $\mathbb{C}_p^\flat$ is endowed with the structure of an algebraically closed non-archimedean field of characteristic p with respect to the addition defined by:

$$(a^{(i)}) + (b^{(i)}) := (c^{(i)}), \qquad c^{(i)} := \lim_{n \to \infty} (a^{(i+n)} + b^{(i+n)})^{p^n}.$$

We can take $\varpi = (p^{1/p^n})$ for some choice of compatible p-primary roots of p. The field $\mathbb{C}_p^\flat$ is called the *tilt* of $\mathbb{C}_p$. One can show that $\mathbb{C}_p^\flat$ is non-canonically isomorphic to $\widehat{\overline{\mathbb{F}_p((t))}}$.

In what follows we consider the Fargues-Fontaine curve associated with the field $C = \mathbb{C}_p^\flat$. There is a natural homomorphism

$$\theta : W(\mathcal{O}_{\mathbb{C}_p^\flat}) \longrightarrow \mathcal{O}_{\mathbb{C}_p} \qquad \sum_{i \geqslant 0} [a_i] p^i \longmapsto \sum_{i \geqslant 0} a_i^{(0)} p^i .$$

This induces a morphism of adic spaces

$$\iota_\infty : \mathrm{Spa}(\mathbb{C}_p) \longrightarrow X_{FF},$$

that is, a closed point ∞ on the Fargues-Fontaine curve.

The completed local ring $\widehat{\mathcal{O}}_{X_{FF},\infty}$ is isomorphic to the ring B_{dR}^+, which is a complete discrete valuation ring with residue field $\mathbb{C}_p$. Its fraction field $B_{\mathrm{dR}} = B_{\mathrm{dR}}^+[\frac{1}{t}]$ is Fontaine's period ring [20], where $t \in B_{\mathrm{dR}}^+$ is a uniformizer.

More generally, the other rigid points of X_{FF} classify all other possible choices of "untilts" of $\mathbb{C}_p^\flat$ or equivalently, endomorphisms of $\mathbb{C}_p^\flat$ modulo Frobenius [35, Thm. 1.4.13].

Now we come back to p-divisible groups. The following result is by Scholze and Weinstein [50, Thm. 6.2.1]:

Theorem 2.61. *Let $\mathbb{X}$ be a p-divisible group over $\bar{\mathbb{F}}_p$ of height n and dimension d. Let (W, φ_W) be the isocrystal associated of $\mathbb{X}$ and let $\mathcal{E}$ be the associated rank n vector bundle on X_{FF}. Then $\mathcal{M}_{\mathbb{X},\infty}(\mathbb{C}_p)$ coincides canonically with the set of injective morphisms $f \colon \mathcal{O}_{X_{FF}}^n \hookrightarrow \mathcal{E}$ such that $\mathrm{coker}(f)$ is isomorphic to $\iota_{\infty*} V$ for some $\mathbb{C}_p$-vector space V (which will necessarily be of dimension d).*

The property of being isomorphic to $\iota_{\infty*} V$ for some $\mathbb{C}_p$-vector space V can be referred to as "being scheme-theoretically supported at ∞."

The maps f of the statement can be interpreted as *modifications of $\mathcal{O}_{X_{FF}}^n$ into $\mathcal{E}$ at ∞ of type $\bar{\mu}$*, where $\bar{\mu}$ is the conjugacy class of the cocharacter of $\boldsymbol{GL}_n$ given by the formula

$$\lambda \longmapsto \begin{pmatrix} \lambda & & & & & \\ & \ddots & & & & \\ & & \lambda & & & \\ & & & 1 & & \\ & & & & \ddots & \\ & & & & & 1 \end{pmatrix} \qquad (2.5)$$

where the upper-left block is $d \times d$ and the lower-right one is $(n-d) \times (n-d)$.

2.6.3 Local Shimura Varieties

As in previous sections, let G be a reductive group over $\mathbb{Q}_p$, let $\bar{\mu}$ be a conjugacy class of minuscule cocharacters, and let b be an element of $G(L)$ modulo σ-conjugation. Note that though set $G(L)$ is uncountable, its quotient $B(G) := G(L)/_{\sigma\text{-conj}}$ is actually a countable set (by Kottwitz) which admits a combinatorial description.

We take a Tannakian approach in order to define G-bundles on the Fargues-Fontaine curve. By $\mathrm{Rep}_{\mathbb{Q}_p} G$ denote the category of finite-dimensional representations of G over $\mathbb{Q}_p$.

Definition 2.62. A *G-bundle on the curve X_{FF}* is an exact tensor functor from $\mathrm{Rep}_{\mathbb{Q}_p} G$ to the category of vector bundles on X_{FF}.

The following result of Fargues [16] is a generalization of Theorem 2.60.

Theorem 2.63. *There exists a canonical bijection*

$$B(G) \xleftrightarrow{\ 1:1\ } \{G\text{-bundles on } X_{FF}\}/_{\simeq} \, .$$

Denote the bijection from Theorem 2.63 by $b \mapsto \mathcal{E}_b$. The bijection sends the class of the neutral element $1 \in G(L)$ to the trivial G-bundle $\mathcal{E}_1$.

Furthermore, one generalizes Theorem 2.61 as follows: there is a canonical identification between the set of $\mathbb{C}_p$-points on the infinite level local Shimura variety $\mathcal{M}_{(G,b,\bar{\mu}),\infty}(\mathbb{C}_p)$ and the set of modifications from $\mathcal{E}_1$ to $\mathcal{E}_b$ at ∞ of type $\bar{\mu}$.

We now want to investigate the case of not necessarily minuscule cocharacters. To this end, we attempt to better understand non-minuscule modifications. We first make the following heuristic definition.

Definition 2.64. The *B_{dR}-affine Grassmannian* "is" the quotient $Gr_G = G(B_{\mathrm{dR}})/G(B_{\mathrm{dR}}^+)$. An *open Schubert cell* is the $G(B_{\mathrm{dR}}^+)$-orbit $G(B_{\mathrm{dR}}^+)\sigma G(B_{\mathrm{dR}}^+)$ of an element σ of Gr_G.

We need to make sense of this quotient as a space of some kind. This is indeed possible by using diamonds, which are algebraic spaces with respect to the pro-étale topology on perfectoid spaces [49, Def. 9.2.2], as we will see at the end of this section.

Example 2.65. All smooth rigid analytic varieties are pro-étale locally perfectoid and therefore can be associated to a diamond. For example, $\mathbb{G}_m$ admits a pro-étale cover

$$\cdots \longrightarrow \mathbb{G}_m \overset{T \mapsto T^p}{\Longrightarrow} \mathbb{G}_m \overset{T \mapsto T^p}{\Longrightarrow} \mathbb{G}_m$$

which at infinite level is a perfectoid space $\tilde{\mathbb{G}}_m$. This space can be thought of as a sort of "universal cover" of $\mathbb{G}_m$ (yet it has many nontrivial étale covers, related, for example, to the extraction of ℓ-th root of unity). Thus $\mathbb{G}_m$ is a diamond, not being a perfectoid.

The following is well-known:

Proposition 2.66. *Pick a dominant representative μ for each conjugacy class $\bar{\mu}$. We have the following decomposition:*

$$G(B_{\mathrm{dR}}^+) \backslash G(B_{\mathrm{dR}}) / G(B_{\mathrm{dR}}^+) = \bigsqcup_{\bar{\mu}} G(B_{\mathrm{dR}}^+) \mu(t) G(B_{\mathrm{dR}}^+),$$

where, as above, $t \in B_{\mathrm{dR}}^+$ is a uniformizer.

In other words, open Schubert cells are parametrized by conjugacy classes of cocharacters. We denote each one by $Gr_{G,\bar{\mu}}$.

Example 2.67. Take $G = \boldsymbol{GL}_n$. In this case, the Grassmannian parametrizes B_{dR}^+-lattices in B_{dR}^n. Now take μ equal to the cocharacter $\lambda \mapsto \mathrm{diag}$ $(\lambda, \ldots, \lambda, 1, \ldots, 1)$ with d terms equal to λ and $(n - d)$ terms equal to 1 as in (2.5). The associated Schubert cell classifies lattices $\Lambda \subseteq B_{\mathrm{dR}}^n$ such that $t(B_{\mathrm{dR}}^+)^n \subseteq \Lambda \subseteq (B_{\mathrm{dR}}^+)^n$ and $\dim_{\mathbb{C}_p}((B_{\mathrm{dR}}^+)^n / \Lambda) = d$.

Indeed, defining Λ is equivalent to giving its image in $(B_{\mathrm{dR}}^+)^n / t \simeq \mathbb{C}_p^n$, which is in turn equivalent to giving a d-dimensional quotient of $\mathbb{C}_p^n$. This Schubert cell is then just $Gr(d, n)$.

For an arbitrary minuscule cocharacter, one has the following generalization due to Caraiani and Scholze [6].

Theorem 2.68. *If $\bar{\mu}$ is minuscule, then the Schubert cell $Gr_{G,\bar{\mu}}$ is "given by" the flag variety $\mathcal{F}l_{\bar{\mu}}$ associated to $\bar{\mu}$.*

The non-minuscule Schubert cells are more complicated to describe. In this situation, we can have quotients of the form B_{dR}^+ / t^r for some $r > 1$. They are no longer isomorphic to $\mathbb{C}_p$ and hence the classical tools of linear algebra are no longer sufficient.

On the other hand, such quotients still define iterated extension of $\mathbb{C}_p$ by itself. Geometrically, they are successive extensions of $\mathbb{A}^1$ by itself. For example, when $r = 2$ these will be spaces over $\mathbb{A}^1$ with fibers isomorphic to $\mathbb{A}^1$ itself. The desired space does not exist as a rigid analytic space, but it makes sense as a diamond. More precisely, Scholze [49, Sect. 21] proved the following:

Theorem 2.69. *The closure of each open Schubert cell is a diamond. In particular, Gr_G is an ind-diamond.*

Thus, in the world of diamonds, we have variants of local Shimura varieties for the non-minuscule case as well. The nice feature of diamonds is that we can define the notions of étale site and étale cohomology on them.

2.6.4 Hecke Correspondences

We denote by $\mathrm{Perv}_{G(B_{\mathrm{dR}}^+)}(Gr_G/\mathbb{Q}_p)$ and $\mathrm{Perv}_{G(B_{\mathrm{dR}}^+)}(Gr_G/\mathbb{C}_p)$ the categories of $G(B_{\mathrm{dR}}^+)$-equivariant perverse ℓ-adic sheaves on Gr_G over $\mathbb{Q}_p$ and $\mathbb{C}_p$, respectively. One hopes that there exists a geometric Satake equivalence of the following form:

$$\mathrm{Perv}_{G(B_{\mathrm{dR}}^+)}(Gr_G/\mathbb{C}_p) \simeq \mathrm{Rep}(^L G^\circ)$$

which descends to an equivalence

$$\mathrm{Perv}_{G(B_{\mathrm{dR}}^+)}(Gr_G/\mathbb{Q}_p) \simeq \mathrm{Rep}(^L G)$$

such that the highest weight representation $V_{\bar\mu}$ of $^L G$ defined by $\bar\mu$ corresponds to the intersection complex $\mathcal{IC}_{\bar\mu} \in \mathrm{Perv}_{G(B_{\mathrm{dR}}^+)}(Gr_G/\mathbb{Q}_p)$ of the closed Schubert cell attached to $\bar\mu$.

The strategy of the proof consists of an adaptation of the classical versions of the Satake equivalence. What is missing at the moment is the development of the notions of ℓ-adic sheaves, the six functors formalism, the proper and smooth base change results, etc. As a matter of fact, there should not be any technical problem in this direction.

Definition 2.70.

(i) We let Bun_G be the stack of G-bundles on the Fargues-Fontaine curve. It is smooth locally a diamond (that is, an "analytic Artin stack").

(ii) We let $\mathrm{Hecke}_{\bar\mu}$ be the stack parametrizing $(\mathcal{E}, \mathcal{E}', x, \Phi)$ where $\mathcal{E}, \mathcal{E}'$ are G-bundles on X_{FF}, x is a point on X_{FF}, and Φ is a modification from $\mathcal{E}$ to $\mathcal{E}'$ at x of type $\bar\mu$.

(iii) We let X_{FF}^{abs} be the absolute Fargues-Fontaine curve, parametrizing points in X_{FF}.[5]

---

[5]More precisely: as a diamond it coincides with the quotient of the field $L = W(\bar{\mathbb{F}}_p)[\frac{1}{p}]$ by the action of Frobenius.

For any conjugacy class of cocharacters $\bar{\mu}$ we can define a *Hecke correspondence*

which induces a functor (a *Hecke operator*) at the level of derived categories:

$$T_{\bar{\mu}} : D(\mathrm{Bun}_G, \bar{\mathbb{Q}}_\ell) \longrightarrow D(\mathrm{Bun}_G \times X_{FF}^{abs}, \bar{\mathbb{Q}}_\ell)$$
$$\mathcal{F} \longmapsto R\pi_{2*}(\pi_1^* \mathcal{F} \otimes \mathcal{IC}_{\bar{\mu}}).$$

We remark that the Hecke operators encode the cohomology of the local Shimura curve. Indeed, the latter can be obtained by the following cartesian square

$$
\begin{array}{ccc}
\mathcal{M}_{(G,b,\bar{\mu}),\infty} & \longrightarrow & \{\mathcal{E}_1\} \times \{\mathcal{E}_b\} \\
\downarrow & & \downarrow \\
\mathrm{Hecke}_{\bar{\mu}} & \longrightarrow & \mathrm{Bun}_G \times \mathrm{Bun}_G.
\end{array}
$$

There is a notion of *basic* elements of $B(G)$ (see, e.g., [15, end of p. 20]). For example, if $G = \boldsymbol{GL}_n$, then an element $b \in B(G)$ is basic if and only if it has only one slope.

Theorem 2.71 ([15, Theorem 2.26]). *The stack of semistable bundles* Bun_G^{ss} *is an open substack of* Bun_G *and decomposes into*

$$\mathrm{Bun}_G^{ss} \simeq \bigsqcup_{b \in B(G) \text{ basic}} [*/J_b(\mathbb{Q}_p)]$$

where $*$ *is the point-stack and* $J_b(\mathbb{Q}_p)$ *is the* σ*-centralizer of* b.

The stack $[*/G(\mathbb{Q}_p)]$ is an open substack of Bun_G^{ss} corresponding to trivial G-bundle with $b = 1$ and $J_b(\mathbb{Q}_p) = G(\mathbb{Q}_p)$. On the other hand, smooth representations of $G(\mathbb{Q}_p)$ correspond to ℓ-adic sheaves over $[*/G(\mathbb{Q}_p)]$. We would like to have a similar geometrical description of *admissible* representations, that is, representations that are reflexible or, explicitly, isomorphic to their double duals.

One hopes for the following:

(i) There is a notion of constructible sheaves over diamonds and there is an equivalence between constructible sheaves (possibly with infinite-dimensional stalks) over $[*/G(\mathbb{Q}_p)]$ and admissible representations of $G(\mathbb{Q}_p)$.

(ii) The pushforward under a proper map and the pullback under a smooth map preserve constructible objects.

(iii) The Hecke operators $T_{\bar\mu}$ preserve constructibility, and factor as follows:

$$T_{\bar\mu} : D_{constr}(\mathrm{Bun}_G, \bar{\mathbb{Q}}_\ell) \longrightarrow D_{constr, W_{\mathbb{Q}_p}}(\mathrm{Bun}_G, \bar{\mathbb{Q}}_\ell)$$

$$D_{constr}(\mathrm{Bun}_G \times X^{abs}_{FF}, \bar{\mathbb{Q}}_\ell),$$

where D_{constr} is the subcategory of constructible objects, and the upper-right category is the category of $W_{\mathbb{Q}_p}$-equivariant objects.

2.6.5 Determining the L-parameters

Fix the following data $\Sigma = (n, V_i, w_i, \alpha, \beta)$ where:

- n is a positive integer;
- $V_1, \ldots, V_n$ are n representations of the group ^LG;
- $w_1, \ldots, w_n$ are n elements of $W_{\mathbb{Q}_p}$;
- α and β are two ^LG-equivariant maps

$$\alpha : \mathbb{1} \longrightarrow V_1 \otimes \ldots \otimes V_n, \qquad \beta : V_1 \otimes \ldots \otimes V_n \longrightarrow \mathbb{1},$$

where $\mathbb{1}$ is the trivial representation of $W^n_{\mathbb{Q}_p}$.

We remark that the Satake equivalence tells us that the representations V_i parametrize some Hecke operators T_{V_i}.

We want to show that we can associate to such data a scalar $\lambda(\pi, \Sigma)$ in $\bar{\mathbb{Q}}_\ell$ via an irreducible, admissible representation π of G.

First, we can associate to π a sheaf $\mathcal{F}_\pi$ on $[*/G(\mathbb{Q}_p)]$ which is embedded in Bun_G via an open immersion $j : [*/G(\mathbb{Q}_p)] \hookrightarrow \mathrm{Bun}_G$. We can then consider the sheaf $j_! \mathcal{F}_\pi$ on Bun_G.

We expect that the image of $j_! \mathcal{F}_\pi$ via a Hecke operator is $W_{\mathbb{Q}_p}$-equivariant. In particular, $T_{\bar\mu} j_! \mathcal{F}_\pi$ comes with an action of $W_{\mathbb{Q}_p}$. More generally, every time we compose n Hecke operators, the image of $j_! \mathcal{F}_\pi$ will carry a canonical action of $W_{\mathbb{Q}_p} \times \ldots \times W_{\mathbb{Q}_p}$. The compatibility between these n actions can be obtained by means of the Satake equivalence.

Consider now the element $(T_{V_1} \circ \ldots \circ T_{V_n}) j_! \mathcal{F}_\pi$ and restrict it to the stack $[*/G(\mathbb{Q}_p)]$. By the above, this defines a representation H_π of $G(\mathbb{Q}_p) \times W^n_{\mathbb{Q}_p}$. We let H'_π be a representation of $G(\mathbb{Q}_p) \times W_{\mathbb{Q}_p}$ obtained as the restriction of H_π along the diagonal embedding $\Delta : W_{\mathbb{Q}_p} \to W^n_{\mathbb{Q}_p}$.

By construction, the map α determines a $G(\mathbb{Q}_p) \times W_{\mathbb{Q}_p}$-equivariant morphism $\alpha' : \pi \to H'_\pi$ and similarly the map β determines $\beta' : H'_\pi \to \pi$. On the other hand, the element $(w_1, \ldots, w_n) \in W^n_{\mathbb{Q}_p}$ induces an endomorphism w' of H_π, considered as a $G(\mathbb{Q}_p)$-representation.

As a whole we obtain by composition the map $\beta' \circ w' \circ \alpha'$ which is a $G(\mathbb{Q}_p)$-equivariant endomorphism $\pi \to \pi$. By Schur's lemma, any such endomorphism is actually defined by a scalar $\lambda = \lambda(\pi, \Sigma)$, as desired.

We now follow a similar procedure, but this time starting with a semisimple representation $\varphi \colon W_{\mathbb{Q}_p} \to {}^L G$ instead of π. It suffices to consider the composition

$$\mathbb{1} \xrightarrow{\alpha} (V_1 \circ \varphi) \otimes \ldots \otimes (V_n \circ \varphi_n) \xrightarrow{w} (V_1 \circ \varphi) \otimes \ldots \otimes (V_n \circ \varphi_n) \xrightarrow{\beta} \mathbb{1}$$

where the middle map is induced by the elements $w_1, \ldots, w_n$. Also in this case, we obtain an endomorphism of the trivial representation $\mathbb{1}$ on $W_{\mathbb{Q}_p}$, that is, a scalar $\mu = \mu(\varphi, \Sigma)$.

The functions $\lambda_\pi = \lambda(\pi, \cdot)$ and $\mu_\varphi = \mu(\varphi, \cdot)$ satisfy many properties related to the so-called *excursion operators* and can actually be completely classified, as proved by V. Lafforgue [37]:

Theorem 2.72. *For any admissible irreducible representation π of $G(\mathbb{Q}_p)$ there exists a unique semisimple representation $\varphi = \varphi_\pi \colon W_{\mathbb{Q}_p} \to {}^L G$ such that $\lambda_\pi = \mu_\varphi$.*

We have therefore obtained the association $\pi \mapsto \varphi_\pi$ that we envisaged. We remark that, in some ways, the element φ_π was obtained as an "eigenvalue" of $j_! \mathcal{F}_\pi$.

BIBLIOGRAPHY

[1] Y. André: *Period mappings and differential equations. From $\mathbb{C}$ to $\mathbb{C}_p$*, MSJ Memoirs **12**, Mathematical Society of Japan (2003).

[2] P. Berthelot, A. Ogus: *Notes on Crystalline Cohomology*, Mathematical Notes **21**, Princeton University Press (1978).

[3] A. Borel: *Automorphic L-functions*, Borel, Casselman (eds.), Automorphic Forms, Representations, and *L*-functions, Proceedings of Symposia in Pure Mathematics **33/2**.

[4] S. Bosch, U. Güntzer, R. Remmert: *Non-Archimedean analysis*, Grundlehren der Mathematischen Wissenschaften **261**, Springer-Verlag, Berlin (1984).

[5] O. Brinon, B. Conrad: *Notes on p-adic Hodge theory*, available online.

[6] A. Caraiani, P. Scholze: *On the generic part of the cohomology of compact unitary Shimura varieties*, preprint (2016); available at arXiv:1511.02418.

[7] H. Carayol: *Nonabelian Lubin-Tate theory*, Automorphic forms, Shimura varieties, and *L*-functions, vol. II (Ann Arbor, MI, 1988), Perspect. Math. **11**, Academic Press, Boston, MA (1990), 15-39.

[8] J.W.S. Cassels, A. Fröhlich (eds.): *Algebraic Number Theory*, Academic Press (1967).

[9] P. Deligne, D. Kazhdan, M.-F. Vigneras: *Representations des algebres centrales simples p-adiques*, Representations of reductive groups over a local field, Travaux en Cours, Hermann, Paris (1984), 33-117.

[10] M. Demazure: *Lectures on p-Divisible Groups*, Lecture Notes in Math. **302**, Springer-Verlag (1972).

[11] V. G. Drinfeld: *Elliptic modules*, Mathematics of the USSR-Sbornik **23**:4 (1974), 561-592.

[12] V. G. Drinfeld: *Coverings of p-adic symmetric domains*, Functional Analysis and Its Applications **10**:2 (1976), 107-115.

[13] G. Faltings: *p-adic Hodge theory*, J. Amer. Math. Soc. **1** (1988), 255-299.

[14] G. Faltings: *A relation between two moduli spaces studied by V.G. Drinfeld*, Contemporary Mathematics **300** (2002).

[15] L. Fargues: *Geometrization of the local Langlands correspondence: an overview*, preprint (2016); available at webusers.imj-prg.fr/~laurent .fargues/Geometrization_review.pdf.

[16] L. Fargues: *G-torseurs en thorie de Hodge p-adique*, preprint (2015); available at webusers.imj-prg.fr/~laurent.fargues/Gtorseurs.pdf.

[17] L. Fargues, J.-M. Fontaine: *Courbes et fibrés vectoriels en théorie de Hodge p-adique*, preprint (2015); available at webusers.imj-prg.fr/~laurent .fargues/Courbe_fichier_principal.pdf.

[18] L. Fargues: *Cohomologie des espaces de modules de groupes p-divisible et correspondances de Langlands locales,* Fargues, Mantovan, Variétés de Shimura, espaces de Rapoport–Zink et correspondances de Langlands locales, Astérisque **291** (2004).

[19] L. Fargues, A. Genestier, V. Lafforgue: *L'isomorphisme entre les tours de Lubin–Tate et de Drinfeld*, Progress in Mathematics **262**, Birkhäuser.

[20] J.-M. Fontaine: *Groupes p-divisibles sur les corps locaux*, Astérisque **47-48** (1977).

[21] J. Fresnel, M. van der Put: *Rigid analytic geometry and its applications*, Progress in Mathematics **218**, Birkhäuser.

[22] B. Gross, M. Hopkins: *Equivariant vector bundles on the Lubin–Tate moduli space*, E. Friedlander, M. Mahowald (eds.), Papers from the Conference on Connections between Topology and Representation Theory held at Northwestern University, Evanston, Illinois, May 1-5, 1992, Contemporary Mathematics **158**, American Mathematical Society (1994).

[23] A. Grothendieck: *Groupes de Barsotti-Tate et cristaux de Dieudonné*, Les Presses de L'Université de Montréal (1974).

[24] A. Grothendieck: *Groupes de Barsotti-Tate et cristaux*, ICM Proceedings (1970).

[25] M. Harris, R. Taylor: *The geometry and cohomology of some simple Shimura varieties*, Annals of Mathematics Studies **151** (2001).

[26] M. Hazewinkel: *Formal groups and applications*, Pure and Applied Mathematics **78**, Academic Press, New York-London (1978).

[27] G. Henniart: *Une preuve simple des conjectures de Langlands pour GL(n) sur un corps p-adique*, Inventiones Mathematicae **139**:2 (2000), 439-455.

[28] R. Huber: *Continuous valuations*, Math. Z. **212**:3 (1993), 455-477.

[29] R. Huber: *A generalization of formal schemes and rigid analytic varieties*, Math. Z. **217**:4 (1994), 513-551.

[30] R. Huber: *Étale cohomology of rigid analytic varieties and adic spaces*, Aspects of Mathematics, E30, Friedr. Vieweg & Sohn, Braunschweig (1996).

[31] L. Illusie: *Complexe Cotangent et Déformations II*, Lecture Notes in Math. **283**, Springer-Verlag (1972).

[32] L. Illusie: *Déformations de groupes de Barsotti-Tate*, Séminaire sur les pinceaux arithmétiques, Astérisque **127** (1985).

[33] H. Jacquet, R. P. Langlands: *Automorphic forms on GL(2)*, Lecture Notes in Mathematics **114**, Springer-Verlag, Berlin-New York (1970).

[34] N. Katz: *Serre-Tate Local Moduli*, Surfaces Algébriques, Lecture Notes in Math. **868**, Springer-Verlag (1981).

[35] K. Kedlaya: *New methods for (φ, Γ)-modules*, Research in the Mathematical Sciences **2**:20 (Robert Coleman memorial issue) (2015).

[36] R. Kottwitz: *Isocrystals with additional structure II*, Compositio Mathematica **109** (1997), 255-339

[37] V. Lafforgue: *Introduction aux chtoucas pour les groupes rductifs et la paramtrisation de Langlands globale*, preprint (2015); available at arXiv: 1404.3998.

[38] J. Lubin, J. Tate: *Formal moduli for one-parameter formal Lie groups*, Bulletin de la S.M.F. **94** (1966), 49-59.

[39] Y. Manin: *The theory of commutative formal groups over fields of finite characteristic*, Russ. Math. Surv. **18**:1 (1963).

[40] B. Mazur, W. Messing: *Universal extensions and one dimensional crystalline cohomology*, Lecture Notes in Math. **370**, Springer-Verlag (1974).

[41] W. Messing: *The Crystals Associated with Barsotti–Tate schemes: with Applications to Abelian Schemes*, Lecture Notes in Math. **264**, Springer-Verlag (1972).

[42] M. Rapoport: *Non-Archimedean period domains*, ICM Proceedings (1994).

[43] M. Rapoport: *A guide to the reduction modulo p of Shimura varieties*, Automorphic forms I, Astérisque **298** (2005).

[44] M. Rapoport, E. Viehmann: *Towards a theory of local Shimura varieties*, Münster Journal of Mathematics **7** (2014), 273-326.

[45] M. Rapoport, Th. Zink: *Period Spaces for p-divisible Groups*, Annals of Mathematics Studies **141**, Princeton University Press (1996).

[46] J. D. Rogawski: *Representations of GL(n) and division algebras over a p-adic field*, Duke Mathematical Journal **50**:1 (1983), 161–196.

[47] S. Shatz: *Group Schemes, Formal Groups and p-Divisible Groups*, Cornell, Silverman (eds.), Arithmetic Geometry, Springer (1986).

[48] P. Scholze: *Perfectoid spaces*, Publ. Math. Inst. Hautes Études Sci. **116** (2012), 245-313.

[49] *Peter Scholze's lectures on p-adic geometry*, taken by Jared Weinstein (2014); available at math.bu.edu/people/jsweinst/Math274/Scholze Lectures.pdf.

[50] P. Scholze, J. Weinstein: *Moduli of p-divisible groups*, Camb. J. Math. **1**:2 (2013), 145-237.

[51] J. Tate: *p-divisible groups*, Springer (ed.), Proceedings of a Conference on Local Fields (Driebergen 1966), Springer (1967).

[52] T. Wedhorn: *Adic spaces*, preprint (2012); available at math.stanford.edu /~conrad/Perfseminar/refs/wedhornadic.pdf.

[53] *Expository lecture notes from the Arbeitsgemeinschaft on "Geometric Langlands,"* taken by Tony Feng (2016); available at web.stanford.edu /~tonyfeng/Arbeitsgemeinschaft2016.html.

Chapter Three

Hyperelliptic Continued Fractions and Generalized Jacobians: Minicourse Given by Umberto Zannier

Laura Capuano, Peter Jossen, Christina Karolus, and Francesco Veneziano

These are notes from the minicourse given by Umberto Zannier (Scuola Normale Superiore di Pisa). The notes were worked out by Laura Capuano, Peter Jossen,[1] Christina Karolus, and Francesco Veneziano. Most of the material of these lectures, except for the numerical examples which were added by us, is already available in [45]. The authors wish to thank Umberto Zannier for the lively discussions in Alpbach, and Olaf Merkert for providing computations of the examples 3.17, 3.28, 3.29, 3.33, and 3.25.

3.1   INTRODUCTION AND SOME HISTORY

Let d be an integer. The Pell equation, bearing John Pell's (1611–1685) name somewhat by mistake, is the Diophantine equation

$$x^2 - dy^2 = 1$$

to be solved in integers x and y. This equation was studied by Indian, and later by Arabic and Greek, mathematicians (see, for example, [16] for some history on the problem). From a modern point of view, solutions (x, y) of the Pell equation correspond to units $x + y\sqrt{d}$ of norm 1 in the ring $\mathbb{Z}[\sqrt{d}]$. One reason why ancient mathematicians were interested in the Pell equation is that a solution (x, y) of the Pell equation with large x and y provides a good rational approximation to the square root of d, as

$$d = \frac{x^2 - 1}{y^2} \simeq \left(\frac{x}{y}\right)^2.$$

---

[1] P. Jossen served as group leader of the working group "Minicourse Zannier."

For instance, Baudhayana (a Vedic priest who lived around 800 BC) discovered that $(x, y) = (17, 12)$ and $(x, y) = (577, 408)$ are solutions for the Pell equation with $d = 2$, and that $17/12$ and $577/408$ are close approximations to $\sqrt{2}$. In fact,

$$\frac{577}{408} = 1.41421568627\ldots \qquad \text{and} \qquad \sqrt{2} = 1.41421356237\ldots$$

Methods to construct new, larger solutions of the Pell equation from a given solution were already known to the Indian mathematician and astronomer Brahmagupta in the seventh century. The fact that for every nonsquare $d > 0$ the Pell equation has one (hence infinitely many) nontrivial solutions is a result attributed to Lagrange. Long before him, Wallis and Euler described methods finding solutions of the Pell equation, although Lagrange was the first to show that the method actually works in any case. Euler's method involves continued fractions. For example, to solve the equation $x^2 - 3y^2 = 1$, we can write

$$\sqrt{3} = 1 + \cfrac{1}{1 + \cfrac{1}{2 + \sqrt{3}}} = 1 + \cfrac{1}{1 + \cfrac{1}{2 + \cfrac{1}{1 + \cfrac{1}{2 + \cdots}}}},$$

and notice that the continued fraction of $\sqrt{3}$ is periodic. Stopping the continued fraction at various stages yields a sequence of rational approximations to $\sqrt{3}$. These are:

$$1, \frac{2}{1}, \frac{5}{3}, \frac{7}{4}, \frac{19}{11}, \frac{26}{15}, \frac{71}{41}, \frac{97}{56}, \frac{265}{153}, \frac{362}{209}, \frac{989}{571}, \frac{1351}{780}, \frac{3691}{2131}, \frac{5042}{2911}, \frac{13775}{7953}, \ldots$$

In this sequence, solutions (x, y) of the Pell equation $x^2 - 3y^2 = 1$ occur as numerators and denominators, as one stops the continued fraction at even stages:

$$1 = 2^2 - 3 \cdot 1^2 = 7^2 - 3 \cdot 4^2 = 26^2 - 3 \cdot 15^2 = \cdots = 5042^2 - 3 \cdot 2911^2.$$

If we stop at odd stages we solve the equation $x^2 - 3y^2 = -2$. Lagrange's contribution is the statement that for every nonsquare integer $d > 1$, the continued fraction expansion of $\sqrt{d}$ is periodic.

In the 1760s, Euler discovered several polynomial identities for the Pell equation. Among them, for example, the following equality ([11]):

$$(2n^2 + 1)^2 - (n^2 + 1)(2n)^2 = 1. \tag{3.1}$$

More generally, if T_k and U_k denote the Chebyshev polynomials of the first and second kind, the relation

$$T_k(n)^2 - (n^2 - 1)U_{k-1}(n)^2 = 1$$

holds. Euler's identity is the case $k = 2$. Such polynomial solutions to the Pell equation have interesting applications to the problem of computing class numbers of real quadratic number fields (see [20]) but also qualify as interesting for their own sake.

In these notes, we study the polynomial interpretation of the Pell equation

$$x(t)^2 - D(t)y(t)^2 = 1,$$

where $D(t) \in \mathbb{C}[t]$ is a given polynomial with complex coefficients of even degree, to be solved in polynomials $x(t), y(t) \in \mathbb{C}[t]$. This topic was already studied by Abel in 1826, later also by Chebyshev and, more recently, among others by Hellegouarch, van der Poorten, Platonov, Akhiezer, Krichever, McMullen, Masser, Bertrand, and Zannier. Abel was interested in expressing certain integrals in "finite terms": He observed that, if $x(t), y(t) \in \mathbb{C}[t]$ form a nontrivial solution of the Pell equation, then the equality

$$\int \frac{x'(t)}{y(t)\sqrt{D(t)}} dt = \log\left(x(t) + y(t)\sqrt{D(t)}\right)$$

holds. As in the arithmetic case, there is a close connection between continued fractions and the solutions of the Pell equation. Namely, if we expand $\sqrt{D(t)}$ as a Laurent series around ∞ and determine its continued fraction expansion (a procedure we shall explain in more details later), then the following holds.

Theorem 3.1 (Abel, 1826).   *Let $D(t) \in \mathbb{C}[t]$ be a polynomial of even degree, which is not a perfect square. The Pell equation $x(t)^2 - D(t)y(t)^2 = 1$ has a nontrivial solution if and only if the continued fraction expansion of $\sqrt{D(t)}$ is eventually periodic.*

Among the myriad of interesting questions that one may ask about continued fraction expansions of algebraic functions such as $\sqrt{D(t)}$, or of Laurent series in general, we will focus on two. The first concerns the behavior of the solvability of the polynomial Pell equation for families of polynomials. Consider the family of polynomials $D_\lambda(t) \in \mathbb{C}(\lambda)[t]$ depending on a parameter λ, for example, $D_\lambda(t) = t^4 + \lambda t^2 + t + 1$. We may ask for which specializations of the parameter $\lambda \in \mathbb{C}$ the equation

$$x(t)^2 - D_\lambda(t)y(t)^2 = 1$$

has a nontrivial solution. These problems for pencils of nonsquarefree polynomials have been studied by Masser and Zannier in [23] and [19] (see also other results by Bertrand [4] and Schmidt [31] for some nonsquarefree families of $D(t)$ and Barroero-Capuano [2] for some results on the generalized Pell equation). We also point out that these questions are related to problems of *unlikely intersections* in families of Jacobians of hyperelliptic curves (or *generalized Jacobians*, in the nonsquarefree case). For a survey on this, see also [44]. We will discuss this matter in Section 3.6.

The second question concerns the behavior of the partial quotients in the continued fraction expansion of $\sqrt{D(t)}$ in the non-periodic case. Here we will prove that at least the sequence of degrees of the partial quotients is periodic, which is a recent result of Zannier (Theorem 3.30). We will give a proof of this result in Section 3.9.

3.2   THE CONTINUED FRACTION EXPANSION OF REAL NUMBERS

In this section we review several classical definitions and results related to the continued fraction expansion of real numbers, and illustrate them by examples. A good general reference is Khinchin's book [13].

— **3.2.**   Let r be a real number. The continued fraction expansion of r is an expression, either finite or not, of the form

$$r = a_0 + \cfrac{1}{a_1 + \cfrac{1}{a_2 + \cfrac{1}{a_3 + \cdots}}}$$

where the a_n are integers, and are positive for $n \geq 1$. The continued fraction expansion of a given real number r can be obtained as follows. Denote by $\lfloor r \rfloor$ the integral part of r, that is, the largest integer which is smaller or equal to r, so that $0 \leq r - \lfloor r \rfloor < 1$ holds. Set $a_0 = \lfloor r \rfloor$ and $r_0 = r - a_0$, then if $r_n \neq 0$, put $a_{n+1} = \lfloor r_n^{-1} \rfloor$ and $r_{n+1} = r_n^{-1} - a_{n+1}$. If ever $r_n = 0$, which happens if and only if r is a rational number, the procedure stops. The integers $a_0, a_1, \ldots$ are called *partial quotients*. As a matter of notation, we usually denote the continued fraction by

$$r = [a_0; a_1, a_2, a_3, \ldots]. \tag{3.2}$$

Given any finite or infinite sequence of integers $a_0, a_1, a_2 \ldots$, we define two new sequences $\{p_n\}$ and $\{q_n\}$ by setting[2]

$$\begin{cases} p_{n+1} = a_n p_n + p_{n-1}, & p_{-1} = 0 \quad \text{and} \quad p_0 = 1 \\ q_{n+1} = a_n q_n + q_{n-1}, & q_{-1} = 1 \quad \text{and} \quad q_0 = 0. \end{cases} \tag{3.3}$$

A more elegant way of rewriting (3.3) is

$$\begin{pmatrix} a_0 & 1 \\ 1 & 0 \end{pmatrix} \begin{pmatrix} a_1 & 1 \\ 1 & 0 \end{pmatrix} \begin{pmatrix} a_2 & 1 \\ 1 & 0 \end{pmatrix} \cdots \begin{pmatrix} a_n & 1 \\ 1 & 0 \end{pmatrix} = \begin{pmatrix} p_{n+1} & p_n \\ q_{n+1} & q_n \end{pmatrix}$$

---

[2]Several authors use shifted indices, so that their p_n are our p_{n+1}.

from which we obtain the relation $p_n q_{n-1} - q_n p_{n-1} = (-1)^n$. In particular, p_n and q_n are coprime. We have, for instance,

$$
\begin{aligned}
p_0 &= a_0 & p_1 &= a_0 a_1 + 1 & p_2 &= a_0 a_1 a_2 + a_0 + a_2 \\
q_0 &= 1 & q_1 &= a_1 & q_2 &= a_1 a_2 + 1.
\end{aligned}
$$

We may look at p_n and q_n as elements of the ring of polynomials $\mathbb{Z}[a_0, a_1, \ldots]$. The equality

$$
\frac{p_n}{q_n} = [a_0; a_1, a_2, \ldots, a_{n-1}] = a_0 + \cfrac{1}{a_1 + \cfrac{1}{a_2 + \cdots \cfrac{\cdots}{\cfrac{1}{a_{n-1}}}}}
$$

holds in the fraction field of $\mathbb{Z}[a_0, a_1, \ldots]$. In our concrete situation, the ratios p_n/q_n are just rational numbers, called *convergents*, and the meaning of equality (3.2) is that

$$
\lim_{n \to \infty} \frac{p_n}{q_n} = r \tag{3.4}
$$

holds.

— **3.3.** The convergents p_n/q_n obtained from the continued fraction expansion of a real number r are the "best" rational approximations of r; this statement can be made precise in several different ways.

The convergents approximate r better than any other rational number with a smaller denominator: If p_n/q_n is a convergent (so it is automatically a reduced fraction), then the inequality

$$
|p_n - q_n r| < |p - q r|
$$

holds for all rational numbers p/q with $q \le q_n$ and $p/q \ne p_n/q_n$. Vice versa, if p/q is a rational number with the property that the inequality $|p - qr| < |p' - q'r|$ holds for all rational numbers p'/q' with $q' \le q$ and $p'/q' \ne p/q$, then p/q is a convergent.

The convergents also have the property that they approximate the number r with an error very small compared to the denominator q_n:

$$
\left| \frac{p_n}{q_n} - r \right| < \frac{1}{q_n^2}.
$$

The converse of this statement holds up to a factor 2: If p/q is a rational number such that

$$
\left| \frac{p}{q} - r \right| < \frac{1}{2q^2},
$$

then p/q is a convergent of the continued fraction expansion of r.

Example 3.4. Let us compute the continued fraction expansion of $\sqrt{13}$. We have $3 < \sqrt{13} < 4$, so $a_0 = 3$ and $r_0 = \sqrt{13} - 3$. We continue with the algorithm, where at each step we rationalize the denominators:

$$a_1 = \left\lfloor \frac{1}{\sqrt{13}-3} \right\rfloor = \left\lfloor \tfrac{1}{4}(\sqrt{13}+3) \right\rfloor = 1 \qquad r_1 = \tfrac{1}{4}(\sqrt{13}+3) - 1 = \tfrac{1}{4}(\sqrt{13}-1)$$

$$a_2 = \left\lfloor \frac{4}{\sqrt{13}-1} \right\rfloor = \left\lfloor \tfrac{1}{3}(\sqrt{13}+1) \right\rfloor = 1 \qquad r_2 = \tfrac{1}{3}(\sqrt{13}+1) - 1 = \tfrac{1}{3}(\sqrt{13}-2)$$

$$a_3 = \left\lfloor \frac{3}{\sqrt{13}-2} \right\rfloor = \left\lfloor \tfrac{1}{3}(\sqrt{13}+2) \right\rfloor = 1 \qquad r_3 = \tfrac{1}{3}(\sqrt{13}+2) - 1 = \tfrac{1}{3}(\sqrt{13}-1)$$

$$a_4 = \left\lfloor \frac{3}{\sqrt{13}-1} \right\rfloor = \left\lfloor \tfrac{1}{4}(\sqrt{13}+1) \right\rfloor = 1 \qquad r_4 = \tfrac{1}{4}(\sqrt{13}+1) - 1 = \tfrac{1}{4}(\sqrt{13}-3)$$

$$a_5 = \left\lfloor \frac{4}{\sqrt{13}-3} \right\rfloor = \left\lfloor (\sqrt{13}+3) \right\rfloor = 6 \qquad r_5 = (\sqrt{13}+3) - 6 = \sqrt{13} - 3$$

We find $r_5 = r_0$, hence $a_6 = a_1$ and $r_6 = r_1$, and the pattern repeats. The continued fraction expansion of $\sqrt{13}$ is therefore given by

$$\sqrt{13} = [3; \overline{1,1,1,1,6}],$$

where the bar indicates that the pattern of partial quotients $1, 1, 1, 1, 6$ repeats periodically. We compute a few convergents. Starting with $p_0 = a_0 = 3$ and $q_0 = 1$, we find

$$\frac{p_1}{q_1} = \frac{3}{1}, \quad \frac{p_2}{q_2} = \frac{4}{1}, \quad \frac{p_3}{q_3} = \frac{7}{2}, \quad \frac{p_4}{q_4} = \frac{11}{3}, \quad \frac{p_5}{q_5} = \frac{18}{5}, \quad \frac{p_6}{q_6} = \frac{119}{33}, \quad \frac{p_7}{q_7} = \frac{137}{38}, \quad \dots$$

and, with the help of a machine,

$$\frac{p_{101}}{q_{101}} = \frac{6787570465375238075075157060001}{1882533334518107155172472208200},$$

which yields about $65 \simeq \log_{10}(p_{101}) + \log_{10}(q_{101})$ correct decimals of $\sqrt{13}$. We point out that, from a computational point of view, continued fractions are not the optimal tool to approximate square roots (Newton's method, for example, is much faster).

Theorem 3.5 (Euler, Lagrange). *Let r be a real number. The continued fraction expansion of r is eventually periodic if and only if r is an irrational algebraic number of degree 2.*

It was Euler's observation that real numbers with an eventually periodic continued fraction expansion satisfy a quadratic equation with integer coefficients, and Lagrange proved the converse by showing that there are only finitely many possible inner terms r_n—in Example 3.4 it is clear that inner terms are of the shape $\frac{1}{P}(\sqrt{13}+Q)$ for integers P, Q of bounded size. Determining the length of the period of the continued fraction expansion of a quadratic algebraic number is a difficult problem. Denoting by $l(d)$ the length of the period of the continued fraction expansion of $\sqrt{d}$ for a nonsquare integer $d > 0$, estimates such as

$$l(d) \leq \frac{7}{2\pi^2} \sqrt{d} \cdot \log(d) + O(\sqrt{d})$$

as $d \to \infty$ can be proven by making Lagrange's finiteness result effective, as done in [9].

— **3.6.** We have explained in 3.3 how convergents of the continued fraction expansion of a real number r are the best possible approximations of r by rational numbers as one imposes an upper bound for the denominator. How well a real number r can be approximated by rational numbers with bounded denominator is measured by the *irrationality measure* $\mu(r)$ of r, also called *Liouville-Roth constant* of r, which is defined as follows. Let $M(r)$ be the set of those real numbers $\mu \in \mathbb{R}$ for which the inequality

$$0 < \left| r - \frac{p}{q} \right| < \frac{1}{q^\mu}$$

has infinitely many solutions in rational numbers p/q, where p and $q > 0$ are integers. The set $M(r)$ is not empty, for it contains the whole interval $(-\infty, 1)$. The irrationality measure of r is defined by $\mu(r) := \sup M(r)$, which is either a real number or the symbol $+\infty$. The most important theorem about irrationality measures is Roth's theorem.

Theorem 3.7 (Roth, 1955). *Let $r \in \mathbb{R}$ be an irrational, algebraic number. The irrationality measure of r is 2.*

— **3.8.** Roth's theorem, also called the Thue-Siegel-Roth theorem, has a long history, starting with Dirichlet and Liouville. For a real number r, there are two possible regimes for $\mu(r)$:

$$\mu(r) = \begin{cases} = 1 & \text{if and only if } r \text{ is rational,} \\ \geq 2 & \text{if } r \text{ is irrational (exactly equal to 2 if } r \text{ is algebraic).} \end{cases}$$

All algebraic irrational numbers satisfy $\mu(r) = 2$, but there exist also transcendental numbers with irrationality measure equal to two; this can be shown by a simple counting argument, but it is also known that $\mu(e) = 2$ holds. The numbers for which $\mu(r) = +\infty$ are called *Liouville numbers*; an example of a Liouville number is Liouville's constant

$$L = \sum_{n=1}^{\infty} 10^{-n!} = 1.1000100000000000000000100\ldots$$

which served as an explicit example of a transcendental number in 1850, about forty years before Cantor's diagonal argument. The relation between continued fractions and irrationality measures, already established in Paragraph 3.3, is further illustrated by the following proposition.

Proposition 3.9 ([35], Thm. 1). *Let r be a real number with continued fraction expansion $r = [a_0; a_1, a_2, \ldots]$. The irrationality measure $\mu(r)$ of r is given by*

$$\mu(r) = 1 + \limsup_{n \to \infty} \frac{\log(q_{n+1})}{\log(q_n)} = 2 + \limsup_{n \to \infty} \frac{\log(a_n)}{\log(q_n)},$$

where p_n/q_n are the convergents of the continued fraction expansion of r.

Example 3.10. The continued fraction expansion of Liouville's constant starts with

$$L = [0, 9, 11, 99, 1, 10, 9, 999999999999, 1, 8, 10, 1, 99, 11, 9, 999999999999999$$
$$9999999999999999999999999999999999999999999999999999999, \ldots]$$

and these extremely large terms continue to appear. For infinitely many n, the partial quotient a_{n+1} is much larger than q_n, which is a polynomial expression in $a_0, a_1, \ldots, a_n$.

3.3 CONTINUED FRACTIONS IN MORE GENERAL SETTINGS

One may think of several variants of continued fraction expansions, for real or complex numbers, or even in other fields such as the p-adic numbers. Continued fractions in p-adic numbers were studied by Mahler in [17].

— **3.11.** Let us sum up what we needed in 3.2 in order to create a theory of continued fractions. First of all we need a topological field k, so that the limit of convergents (3.4) makes sense. Then we also need a notion of integral part and fractional part of elements of k. What we want is two subsets I and F of k, which satisfy the following properties:

1. For every $r \in k$ there exists a unique $i \in I$ satisfying $r - i \in F$.
2. The subfield $k_0 \subseteq k$ generated by elements of I is dense in k.
3. $r \in F, r \neq 0 \implies r^{-1} \notin F$.

We call integral part of r and denote by $\lfloor r \rfloor$ the unique element $i \in I$ satisfying $r - i \in F$. A sequence of partial quotients for $r \in k$ with respect to the chosen pair of sets (I, F) can be obtained in the usual way. Set $a_0 = \lfloor r \rfloor$ and $r_0 = r - a_0$, then if $r_n \neq 0$, put $a_{n+1} = \lfloor r_n^{-1} \rfloor$ and $r_{n+1} = r_n^{-1} - a_{n+1}$. If some $r_n = 0$, then r belongs to the subfield of k generated by I, and the procedure stops. The sequence of partial quotients

$$[a_0; a_1, a_2, \ldots]$$

satisfies $a_0 \in I$ and $a_n \in I_0 := \{\lfloor f^{-1} \rfloor \mid f \in F, f \neq 0\}$ for $n \geq 1$. Conditions (2) and (3) are necessary for the sequence of convergents of the so-constructed continued fraction expansion to converge to r, but not sufficient. Condition (3) states that 0 is not an element of I_0. In order to guarantee continued fraction expansions to converge, one should probably replace (3) by a stronger condition which states that elements of I_0 are sufficiently far away from 0.

Example 3.12. Consider the field $k = \mathbb{R}$ with set of integral parts $I = \mathbb{Z}$, but with set of fractional parts $F = [-\frac{1}{2}, \frac{1}{2})$. The continued fraction expansion with respect to this choice will have positive or negative partial quotients $a_n \in \mathbb{Z}$ of absolute value ≥ 2. The reader may compute the continued fraction expansion of $\sqrt{13}$ with respect to this choice of fractional parts. Interestingly enough, one finds a periodic pattern, with period length 3, different from the period length 5 in the standard expansion that was given in Example 3.4, that is,

$$\sqrt{13} = [4; \overline{-3, 2, 7}].$$

Here is a list of convergents:

$$4, \frac{11}{3}, \frac{18}{5}, \frac{137}{38}, \frac{393}{109}, \frac{649}{180}, \frac{4936}{1369}, \frac{14159}{3927}, \frac{23382}{6485}, \frac{177833}{49322}, \frac{510117}{141481}, \frac{842401}{233640}, \ldots$$

The numerators and denominators of the convergents p_n/q_n are solutions to $p_n^2 - 13q_n^2 = c$ where c is $3, 4, -1, -3, -4, 1$, depending on the congruence class of n modulo 6. In particular, we find the solutions

$$649^2 - 13 \cdot 180^2 = 1 \qquad\qquad 842401^2 - 13 \cdot 233640^2 = 1$$

of the Pell equation.

Example 3.13. Consider the field $k = \mathbb{C}$ with set of integral parts $I = \mathbb{Z}[i]$ and set of fractional parts $F = [-\frac{1}{2}, \frac{1}{2}) \times [-\frac{1}{2}, \frac{1}{2})i$. This choice yields a theory of continued fractions for complex numbers which extends the continued fractions for real numbers given in example 3.12. For example, we have

$$\sqrt{2 + 3i} = [2 + i; \overline{-3 + i, 4 + 2i}],$$

where the square root is the one which is about $1.67415 + 0.895977i$. The first few convergents are

$$2 + i, \frac{17 + 9i}{10}, \frac{290 + 155i}{173}, \frac{1239663i}{740}, \frac{42358 + 22669i}{25301}, \frac{72407 + 38751i}{43250}, \ldots$$

but, somewhat disappointingly, numerators and denominators of these convergents do not solve the Pell equation for $d = 2 + 3i$. It was rather important that we chose F as we did, and not $F = [0, 1) \times [0, 1)i$. With the latter choice, reciprocals of elements of F may have too small norms for continued fractions to converge. To see what goes wrong, consider with the latter choice for F the expansion of the 12-th root of unity $\exp(2\pi i/12)$. It is $[0; i, i, i, i, \ldots]$ and does not converge.

Example 3.14. Let $\mathbb{Q}_p$ denote the field of p-adic numbers. There is no canonical way to define the continued fractions in this context, as we do not have a canonical definition of "integral part." Our setup here is the same as Ruban's in [29]. Declare the set of fractional parts to be $F = p\mathbb{Z}_p$, and the set of integral parts I to be the set of all sums

$$c_0 + c_1 p^{-1} + c_2 p^{-2} + \cdots + c_n p^{-n},$$

with $c_i \in \{0, 1, 2, \ldots, p-1\}$. Notice also that rational numbers may have infinite continued fraction expansions, for instance, for $p = 3$ we find

$$\tfrac{1}{7} = [1; \tfrac{1}{3}, \tfrac{7}{3}, \tfrac{8}{3}, \tfrac{8}{3}, \tfrac{8}{3}, \tfrac{8}{3}, \ldots].$$

As a more elaborate example, let us compute the continued fraction expansion of $\sqrt{13}$ in $\mathbb{Q}_3$, where $\sqrt{13}$ is the unique element of $\mathbb{Z}_3$ whose square is 13 and whose class modulo 3 is 1 (and not 2). From $16^2 - 256 \equiv 13 \bmod 243 = 3^5$ we get

$$\sqrt{13} = 1 \cdot 3^0 + 2 \cdot 3^1 + 1 \cdot 3^2 + 0 \cdot 3^3 + 0 \cdot 3^4 + \cdots, \tag{3.5}$$

for some remainder in $3^5 \mathbb{Z}_3$. So $a_0 = 1$ and $r_0 = \sqrt{13} - 1$. To compute the 3-adic expansion of r_0^{-1} we complete the square and use $4 \cdot 61 \equiv 1 \bmod 243$:

$$r_0^{-1} = \tfrac{1}{12}(\sqrt{13} + 1) = 2 \cdot 3^{-1} + 0 \cdot 3^0 + 1 \cdot 3^1 + 2 \cdot 3^2 + 0 \cdot 3^3 + \cdots.$$

From this expansion we read off $a_1 = 2 \cdot 3^{-1}$ and $r_1 = \tfrac{1}{12}(\sqrt{13} - 7)$, and proceed with calculating the 3-adic expansion of r_1^{-1} in the same fashion

$$r_1^{-1} = \tfrac{-1}{3}(\sqrt{13} + 7) = 1 \cdot 3^{-1} + 1 \cdot 3^0 + 0 \cdot 3^1 + 2 \cdot 3^2 + 2 \cdot 3^3 + \cdots,$$

hence $a_2 = \tfrac{4}{3}$. Next up we find

$$r_2^{-1} = \tfrac{1}{36}(\sqrt{13} - 11) = 2 \cdot 3^{-2} + 2 \cdot 3^{-1} + 0 \cdot 3^0 + 2 \cdot 3^1 + 2 \cdot 3^2 + \cdots,$$

hence $a_3 = \tfrac{8}{9}$. So far we have computed

$$\sqrt{13} = [1; \tfrac{2}{3}, \tfrac{4}{3}, \tfrac{8}{9}, \ldots]$$

by hand. Here is a machine computation using Sage [36]. We start with

```
(1) sage: R=Zp(3, prec = 1000, print_mode = 'series')
(2) sage: A=sqrt(R(13))
```

so A is the square root of 13 in $\mathbb{Z}_3$ up to precision 3^{1000}. Printing A yields the first 999 terms of the series representation (3.5). To compute the first 100 terms in the continued fraction expansion, we use the following algorithm:

```
(3)  sage: n=100
(4)  sage: fraction=[]
(5)  sage: for i in range(n):
(6)  sage:     v=A.valuation()
(7)  sage:     B=A/3^v
(8)  sage:     C=B.residue(1-v)
(9)  sage:     D=int(C)
(10) sage:     DD=D*3^v
(11) sage:     print DD
(12) sage:     fraction = fraction + [[D , v]]
(13) sage:     A=1/(A-R(D)*3^v)
```

It works as follows. In lines (3) and (4) we choose the number n=100 of iteration steps, and create an empty list named fraction. We need this list only later to compute convergents. Lines (6) to (13) are then repeated n times. In line (6) we assign to v the 3-adic valuation of A, which is zero or a negative integer, and in line (7) scale A to a 3-adic integer B of valuation 0. Then we define C to be the residue modulo 3^{-v+1}, which encodes the first $-v+1$ coefficients in the series expansion of A. Sage [36] sees C as an element of $\mathbb{Z}/3^{-v+1}\mathbb{Z}$, and we need to reconvert C to an integer D and scale back by the power of 3 we divided by in line (7). Now DD is the integral part of the series expansion of A, and we print it. In line (12) we add the pair (D,V) to the list fraction for later use. Finally, in line (13) we subtract from A its integral part and invert. Here is the output:

1	$\frac{2}{3}$	$\frac{4}{3}$	$\frac{8}{9}$	$\frac{5}{3}$	$\frac{4}{3}$	$\frac{5}{9}$	$\frac{2}{3}$	$\frac{5}{3}$	$\frac{8}{3}$	$\frac{16}{9}$	$\frac{7}{3}$	$\frac{5}{9}$	$\frac{76}{27}$	$\frac{8}{3}$	$\frac{8}{3}$	$\frac{1}{3}$	$\frac{7}{3}$	$\frac{43}{27}$	$\frac{7}{3}$
$\frac{64}{27}$	$\frac{536}{243}$	$\frac{8}{3}$	$\frac{5}{3}$	$\frac{8}{3}$	$\frac{4}{3}$	$\frac{4}{9}$	$\frac{26}{9}$	$\frac{4}{3}$	$\frac{25}{9}$	$\frac{50}{243}$	$\frac{1}{3}$	$\frac{5}{3}$	$\frac{1}{9}$	$\frac{5}{3}$	$\frac{25}{9}$	$\frac{8}{3}$	$\frac{7}{3}$	$\frac{1}{3}$	$\frac{1}{3}$
$\frac{4}{3}$	$\frac{2}{3}$	$\frac{7}{3}$	$\frac{58}{27}$	$\frac{8}{3}$	$\frac{5}{3}$	$\frac{4}{3}$	$\frac{2}{3}$	$\frac{1}{27}$	$\frac{7}{9}$	$\frac{4}{9}$	$\frac{4}{3}$	$\frac{5}{3}$	$\frac{34}{27}$	$\frac{2}{3}$	$\frac{5}{3}$	$\frac{5}{3}$	$\frac{7}{3}$	$\frac{16}{9}$	$\frac{4}{9}$
$\frac{2}{3}$	$\frac{73}{27}$	$\frac{8}{3}$	$\frac{4}{3}$	$\frac{43}{27}$	$\frac{7}{3}$	$\frac{2}{3}$	$\frac{7}{3}$	$\frac{2}{3}$	$\frac{203}{81}$	$\frac{5}{3}$	$\frac{10}{9}$	$\frac{10}{9}$	$\frac{7}{3}$	$\frac{5}{3}$	$\frac{8}{3}$	$\frac{59}{27}$	$\frac{2}{3}$	$\frac{5}{3}$	$\frac{8}{3}$
$\frac{8}{3}$	$\frac{14}{9}$	$\frac{2}{3}$	$\frac{23}{9}$	$\frac{23}{9}$	$\frac{2}{3}$	$\frac{7}{3}$	$\frac{20}{9}$	$\frac{2}{3}$	$\frac{2}{3}$	$\frac{8}{3}$	$\frac{4}{3}$	$\frac{5}{9}$	$\frac{2}{3}$	$\frac{7}{3}$	$\frac{1}{3}$	$\frac{20}{9}$	$\frac{5}{3}$	$\frac{4}{3}$	$\frac{569}{243}$

The first fourteen convergents we get from our computation are:

$$1, \frac{5}{2}, \frac{29}{17}, \frac{367}{190}, \frac{2618}{1409}, \frac{13775}{7346}, \frac{139561}{74773}, \frac{651047}{347888}, \frac{4511284}{2412397}, \frac{41949695}{22430168}, \frac{792999788}{424017407}, \frac{6683640281}{3573736385},$$

$$\frac{54829195681}{29317151914}, \frac{5791143460039}{3096521487019}.$$

The last error term here is

$$\sqrt{13} - \tfrac{5791143460039}{3096521487019} = 1 \cdot 3^{39} + 2 \cdot 3^{41} + 2 \cdot 3^{42} + 2 \cdot 3^{43} + 2 \cdot 3^{44}$$
$$+ 1 \cdot 3^{45} + 1 \cdot 3^{46} + \cdots$$

which is very close to $\sqrt{13}$ in $\mathbb{Z}_3$. To compute the convergents, we used the following algorithm in Sage [36]: After resetting the correct value for A in line (14), it computes the convergents using the Euler-Wallis formulas (3.3), and prints the valuation of the difference $\sqrt{13} - p_n/q_n$.

```
(14) sage: A=sqrt(R(13))
(15) sage: p0=R(fraction[0][0])*3^fraction[0][1]
(16) sage: q0=1
(17) sage: q1=R(fraction[1][0])*3^fraction[1][1]
(18) sage: p1=p0*q1+1
(19) sage: for i in range(2,n):
(20) sage:     an=R(fraction[i][0])*3^fraction[i][1]
(21) sage:     pn=an*p1+p0
(22) sage:     qn=an*q1+q0
(23) sage:     p0=p1
(24) sage:     q0=q1
(25) sage:     p1=pn
(26) sage:     q1=qn
(27) sage:     error= A-pn/qn
(28) sage:     print error.valuation()
```

The output reads

$$6, 9, 11, 14, 17, 19, 21, 24, 27, 30, 35, 39, \ldots, 309 \ldots,$$

which gives us a pretty good idea of what the speed of convergence might be. However, the matter of the convergence of Ruban's continued fraction in $\mathbb{Q}_p$ is not a simple one: unlike the real case, it is not true in general that the convergents always provide good approximations.

It is clear that if $r \in \mathbb{Q}_p$ admits a periodic continued fraction expansion, then r must be a quadratic algebraic number over $\mathbb{Q}$. It is easy to engineer examples in which the continued fraction expansion is periodic. For instance,

$$\tfrac{1}{2p}(-1 + \sqrt{4p^2 + 1}) = [0; \tfrac{1}{p}, \tfrac{1}{p}, \tfrac{1}{p}, \tfrac{1}{p}, \ldots]$$

in $\mathbb{Q}_p$ is periodic and indeed the right-hand side solves the following quadratic equation

$$r = \frac{1}{\tfrac{1}{p} + r}.$$

In a very recent work [10], Capuano, Veneziano, and Zannier found an effective criterion to detect periodicity of Ruban's continued fraction of quadratic irrational numbers. In particular, their criterion shows that $\sqrt{13}$ does not have a periodic continued fraction in $\mathbb{Q}_3$.

Example 3.15. Let $k((s))$ be the field of Laurent series in the variable s and coefficients in a field k. Let us declare the set of fractional parts to be Taylor series with zero constant term, and the set of integral parts to be polynomials in s^{-1}. This choice yields a theory of continued fractions for Laurent series. It is the topic of the next section, except that we shall prefer to work with the variable t^{-1} in place of s, so that integral parts become polynomials in t.

3.4   THE CONTINUED FRACTION EXPANSION OF LAURENT SERIES

In this section we describe the continued fraction expansion of Laurent series, and show some analogies with continued fraction expansions of real numbers. Later we will be interested in the continued fraction expansion of square roots of polynomials.

— **3.16.** Let k be a field, and write $k((t^{-1}))$ for the field of Laurent series in the variable t^{-1} and coefficients in k. An element of $k((t^{-1}))$ is a formal series

$$f(t) = \sum_{n=-\infty}^{n_0} c_n t^n$$

with $c_{n_0} \neq 0$, and we call $\nu(f) := -n_0 \in \mathbb{Z}$ the valuation of f. For $f = 0$ we set $\nu(f) = \infty$. The sets $\{f \in k((t^{-1})) \mid \nu(f) \geq n\}$ form a fundamental system of neighborhoods for a topology on $k((t^{-1}))$. Let us write

$$\lfloor f \rfloor = \sum_{n=0}^{n_0} c_n t^n$$

for the *integral* or *polynomial part* of f. We obtain the continued fraction expansion of a Laurent series $f \in k((t))$ as follows. Set $a_0 = \lfloor f \rfloor$ and $f_0(t) = f(t) - a_0(t)$, and then, if $f_n(t) \neq 0$, set

$$a_{n+1}(t) = \lfloor f_n(t)^{-1} \rfloor \qquad \text{and} \qquad f_{n+1}(t) = f_n(t)^{-1} - a_{n+1}(t)$$

recursively for $n \geq 1$. We obtain the continued fraction of f

$$f(t) = a_0 + \cfrac{1}{a_1 + \cfrac{1}{a_2 + \cdots}} = [a_0; a_1, a_2, \ldots] \tag{3.6}$$

with $a_n \in k[t]$ for every n. The *convergents* p_n/q_n of the sequence of polynomials $a_0, a_1, \ldots$ are given, as in the real case, by the recurrence formula (3.3). The meaning of equation (3.6) is that

$$f(t) = \lim_{n \to \infty} \frac{p_n(t)}{q_n(t)}$$

holds, for the topology on $k((t^{-1}))$ induced by the valuation ν.

Example 3.17. Let us compute a few terms of the continued fraction expansion of the exponential function. The polynomial part of

$$\exp(t^{-1}) = 1 + t^{-1} + \tfrac{1}{2}t^{-2} + \tfrac{1}{3!}t^{-3} + \tfrac{1}{4!}t^{-4} + \cdots$$

is the constant polynomial $a_0 = 1$. Subtract a_0 from $\exp(t^{-1})$, invert, and write the resulting Laurent series:

$$\frac{1}{\exp(t^{-1}) - 1} = t - \frac{1}{2} + \frac{t^{-1}}{12} - \frac{t^{-3}}{720} + \frac{t^{-5}}{30240} - \frac{t^{-7}}{1209600} + \cdots.$$

The polynomial part is $a_1 = t - \frac{1}{2}$. Again, subtract a_1, invert, and write the Laurent series:

$$\frac{1}{\frac{1}{\exp(t^{-1})-1} - t + \frac{1}{2}} = \frac{\exp(t^{-1}) - 1}{\frac{1}{2} + t - (t - \frac{1}{2})\exp(t^{-1})}$$

$$= 12t + \frac{t^{-1}}{5} - \frac{t^{-3}}{700} + \frac{t^{-5}}{63000} - \frac{37t^{-7}}{194040000} + \cdots.$$

The integral part, which is the next partial quotient, is thus $a_2 = 12t$. The following table was calculated for us by Olaf Merkert:

n	0	1	2	3	4	5	6	7	8	9	10	11	12
$a_n(t)$	1	$t - \frac{1}{2}$	$12t$	$5t$	$28t$	$9t$	$44t$	$13t$	$60t$	$17t$	$76t$	$21t$	$92t$
	13	14	15	16	17	18	19	20	21	22	23	24	25
	$25t$	$108t$	$29t$	$124t$	$33t$	$140t$	$37t$	$156t$	$41t$	$172t$	$45t$	$188t$	$49t$
	26	27	28	29	30								
	$204t$	$53t$	$220t$	$57t$	$236t$								

We observe, and once we know what we are looking for, it is not hard to prove either, that for $n \geq 2$ the partial fraction a_n is equal to $(2n-1)t$ for odd

n, and $4(2n-1)t$ for even n. Let us compute a few convergents:

$$1, \quad \frac{\frac{1}{2}+t}{-\frac{1}{2}+t}, \quad \frac{1+6t+12t^2}{1-6t+12t^2}, \quad \frac{\frac{1}{2}+6t+30t^2+60t^3}{-\frac{1}{2}+6t-30t^2+60t^3},$$

$$\frac{1+20t+180t^2+840t^3+1680t^4}{1-20t+180t^2-840t^3+1680t^4}$$

The Taylor series expansion at infinity of the convergent of degree 2 reads

$$1+t^{-1}+\frac{t^{-2}}{2}+\frac{t^{-3}}{6}+\frac{t^{-4}}{24}+\frac{t^{-5}}{144}-\frac{t^{-7}}{1728}-\frac{t^{-8}}{3456}-\frac{t^{-9}}{10368}-\frac{t^{-10}}{41472}+\cdots,$$

hence agrees with $\exp(t^{-1})$ up to order $O(t^{-5})$. These are the so-called *Padé approximations* of the function $\exp(t^{-1})$.

— **3.18.**   We may try to link the irrationality measure of a Laurent series $f \in k((t^{-1}))$ with the degrees of the partial quotients a_n in the continued fraction expansion

$$f = [a_0; a_1, a_2, a_3, \ldots]$$

as we did in Proposition 3.9. In view of the recurrence (3.3) and the fact that $\deg(a_n) > 0$ for all $n > 0$, the equalities

$$\deg p_{n+1} = \deg a_n + \deg p_n$$
$$\deg q_{n+1} = \deg a_n + \deg q_n$$

hold, which make it easy to compute the degrees of convergents. The degrees of the partial quotients a_n are connected to the ranks of the so-called *Hankel matrices*, which are associated to the Laurent coefficients of f. In fact, a partial quotient of large degree amounts to the vanishing of several determinants in these matrices. The convergents provide *Padé approximations* to f, which are of importance in transcendence theory and Diophantine approximation.

— **3.19.**   Let us recapitulate briefly what Padé approximations are. In a standard setup, Padé approximations are associated with power series in a variable t instead of Laurent series in t^{-1}. Let

$$f(t) = \sum_{n=0}^{\infty} c_n t^n \qquad \in k[[t]]$$

be a formal power series, and pick two integers $m \geq 0$ and $n \geq 1$. The Padé approximant of f of order (m, n) is the rational function

$$R(t) = \frac{p(t)}{q(t)} = \frac{a_0 + a_1 t + a_2 t^2 + \cdots + a_m t^m}{1 + b_1 t + b_2 t^2 + \cdots + b_n t^n}$$

which agrees with f up to order $m+n$. The requirement $q(0)=1$ determines p and q uniquely. There exist several efficient algorithms to compute Padé approximants. From an elementary point of view, one has to solve the linear system of $m+n+1$ equations

k-th Taylor coefficient of $q(t)f(t) = k$-th Taylor coefficient of $p(t)$

for $0 \le k \le m+n$ in the $m+n+1$ variables $a_0, \ldots, a_m, b_1, \ldots, b_n$, but this is computationally not very efficient. Now, if f is a Laurent series in t^{-1} rather than a Taylor series in t, say

$$f(t) = \sum_{n=-\infty}^{n_0} c_n t^n \qquad \in k((t)),$$

we can still look for rational functions $R = p/q$ with $\deg p \le m$ and $\deg q \le n$ such that $\nu(f - R)$ is as large as possible. Comparing to the Taylor series case, the difference is that we don't need to prescribe a bound on the degree of the numerator p anymore—there is only so much one can do if $\deg q \le n$ is imposed. We may thus define the n-th Padé approximant of $f \in k((t))$ as the rational function $R = p/q$ with $\deg q \le n$ such that $\nu(f - R)$ is maximal. The next proposition is an analogue of the statements in 3.3.

Proposition 3.20. *Let $f \in k((t^{-1}))$ be a Laurent series in the variable t^{-1}, and let $p(t)$ and $q(t)$ be coprime polynomials. Then $p(t) - q(t)f(t) = O(t^{-\deg q - 1})$ holds if and only if p/q is a convergent of the continued fraction expansion of f.*

Proof. See [26], Prop. 2.1. □

— 3.21. The classical theorem of Roth, which we recalled in 3.7, has a function field analogue, due to S. Uchiyama. For a Laurent series $f \in k((t^{-1}))$ we may consider the set $M(f)$ of those real numbers μ for which the inequality

$$\nu\left(f - \frac{p}{q}\right) > \mu \cdot \deg(q)$$

has infinitely many solutions in rational functions $p/q \in k(t)$, where p and q are polynomials. If we define, as Uchiyama does, an absolute value for Laurent series by setting $|f| := c^{-\nu(f)}$ for some fixed real constant $c > 1$, then the above inequality becomes

$$\left| f - \frac{p}{q} \right| < \frac{1}{|q|^\mu},$$

similar to the inequality in 3.6. Again, $M(f)$ contains $(-\infty, 1)$, and we define the irrationality measure of f to be $\mu(f) := \sup M(f)$. Then [37, Thm. 3(i)] states that if f is not rational, then $\mu(f) \ge 2$ holds, while [37, Thm. 2(i)] is an analogue of Roth's theorem.

Theorem 3.22 (Uchiyama).   *Let k be a field of characteristic zero and let $f \in k((t^{-1}))$ be algebraic but not rational over $k(t)$. Then the irrationality measure of f is 2.*

— 3.23.   Let k be a field of characteristic zero and let $f \in k((t^{-1}))$ be algebraic but not rational over $k(t)$. Uchiyama's theorem states that for any $\epsilon > 0$, the inequality

$$\nu\left(f - \frac{p}{q}\right) > (2 + \epsilon) \cdot \deg(q)$$

has only finitely many solutions $p/q \in k(t)$. The possible periodic behavior of the degrees of the partial quotients in the continued fraction expansion of f is related to a stronger version of Uchiyama's theorem, namely a uniform version with $2 \deg q + O(1)$ in place of $(2 + \epsilon) \deg q$. Using Uchiyama's Theorem by J. Wang [41] and later by M. Ru [28], one could achieve such an estimate for algebraic functions of degree ≤ 3 over $\mathbb{C}(t)$.

— 3.24.   Analogously to the irrationality measure for real numbers, we can express the irrationality measure of a Laurent series in terms of its continued fraction expansion. With notations of 3.18, the equalities

$$\mu(f) = 1 + \limsup_{n \to \infty} \frac{\deg q_{n+1}}{\deg q_n} = 2 + \limsup_{n \to \infty} \frac{\deg a_n}{\deg q_n}$$

hold when f has an infinite continued fraction expansion.

Example 3.25.   Let us look at the function field analogue of Liouville's constant, which we introduced in Example 3.10. Set

$$L(t) = \sum_{n=1}^{\infty} t^{-n!} = t^{-1} + t^{-2} + t^{-6} + t^{-24} + t^{-120} + t^{-720} + \cdots.$$

Power series like this go under the name of *lacunary series*, of which Jacobi's theta function series is another example. The continued fraction expansion of L, again computed by Merkert, reads

$a_0 = 0$			
$a_1 = t - 1$	$a_{11} = -t + 1$	$a_{21} = -t^2$	$a_{31} = t - 1$
$a_2 = t + 1$	$a_{12} = -t^{72}$	$a_{22} = -t - 1$	$a_{32} = t + 1$
$a_3 = t^2$	$a_{13} = t - 1$	$a_{23} = -t + 1$	$a_{33} = -t^2$
$a_4 = -t - 1$	$a_{14} = t + 1$	$a_{24} = -t^{480}$	$a_{34} = -t - 1$
$a_5 = -t + 1$	$a_{15} = t^2$	$a_{25} = t - 1$	$a_{35} = -t + 1$
$a_6 = -t^{12}$	$a_{16} = -t - 1$	$a_{26} = t + 1$	$a_{36} = t^{72}$
$a_7 = t - 1$	$a_{17} = -t + 1$	$a_{27} = t^2$	$a_{37} = t - 1$
$a_8 = t + 1$	$a_{18} = t^{12}$	$a_{28} = -t - 1$	$a_{38} = t + 1$
$a_9 = -t^2$	$a_{19} = t - 1$	$a_{29} = -t + 1$	$a_{39} = t^2$
$a_{10} = -t - 1$	$a_{20} = t + 1$	$a_{30} = -t^{12}$	$a_{40} = -t - 1$

and we observe the sporadic large terms a_6, a_{12}, and a_{24} whose degrees are much larger than the degrees of all previous terms combined. It seems safe to conjecture that $\mu(L) = +\infty$.

3.5 PELL EQUATION IN POLYNOMIALS

In this section we take a closer look at the continued fraction expansion of $f(t) = \sqrt{D(t)}$, viewed as a Laurent series in $s = t^{-1}$. As in the case of continued fraction expansions of real numbers, the behavior of the continued fraction expansion of $\sqrt{D(t)}$ is related to the solvability of the polynomial Pell equation $x(t)^2 - D(t)y(t)^2 = 1$.

Definition 3.26. Let k be a field, and let $D(t) \in k[t]$ be a nonconstant polynomial. We say that D is *Pellian* if the Pell equation

$$x(t)^2 - D(t)y(t)^2 = 1 \tag{3.7}$$

has a solution $x(t), y(t) \in k[t]$, with $y \neq 0$.

— **3.27.** The Pell equation can always be solved by $x = \pm 1$ and $y = 0$. We call this the trivial solution. The notion of Pellianity may depend on the arithmetic of the ground field k. We will often stick to algebraically closed fields, or just to $k = \mathbb{C}$. A polynomial $D(t) \in k[t]$ is Pellian if and only if the polynomial $cD(at + b)$ is Pellian for some $a, c \in k^*$ and $b \in k$. Polynomials of odd degree are never Pellian, so from now on we will only consider polynomials of even degree $2d$. The link between the polynomial Pell equation and continued fractions is given by Abel's Theorem 3.1 (see Theorem 3.1 in the Introduction). It says that D is Pellian if and only if the continued fraction expansion of $\sqrt{D(t)}$ is eventually periodic.

Example 3.28. Let us compute the continued fraction expansion of the square root of the polynomial $D(t) = t^2 + 1$ (which is Pellian by Euler's identity (3.1)). Set $s = \frac{1}{t}$. The Laurent series[3]

$$\sqrt{D(s)} = s^{-1}\sqrt{1 + s^2} = s^{-1} + \frac{s}{2} - \frac{s^3}{8} + \frac{s^5}{16} - \frac{5s^7}{128} + \frac{7s^9}{256} - \frac{21s^{11}}{1024} + \cdots$$

has polynomial part $a_0 = s^{-1} = t$. For the next step, we have to compute the Laurent expansion of $(\sqrt{D(s)} - a_0)^{-1}$:

---

[3]Such a Laurent expansion would not exist if D had an odd degree, because then the two roots of $D(t)$ would be interchanged by monodromy around ∞. In other words, if $D(t)$ has odd degree $< 2d - 1$, then the polynomial $s^{2d}D(s)$ has a simple zero at $s = 0$, hence $s^d\sqrt{D(s)}$ does not define an analytic continuation around $s = 0$.

$$(\sqrt{D(s)} - s^{-1})^{-1} = 2s^{-1} + \frac{s}{2} - \frac{s^3}{8} + \frac{s^5}{16} - \frac{5s^7}{128} + \frac{7s^9}{256} - \frac{21s^{11}}{1024} + \cdots.$$

We find $a_1 = 2t$. The remainder $(\sqrt{D(s)} - s^{-1})^{-1} - 2s^{-2}$ is the same as the one obtained in the previous step, and the continued fraction expansion of $\sqrt{D(t)}$ is thus periodic:

$$\sqrt{D(t)} = [t; 2t, 2t, 2t, \ldots].$$

To justify this properly, set $h(t) = \sqrt{D(t)} - t$. We need to show that $h(t)^{-1} - 2t = h(t)$ holds, but this is immediate: Completing the square in the denominator in the left-hand side

$$\frac{1}{\sqrt{t^2 + 1} - t} - 2t = \sqrt{t^2 + 1} - t,$$

the equality is a consequence of Euler's identity (3.1). Therefore, this example is an illustration of Abel's Theorem 3.1. As a corollary, we find the continued fraction expansion of $\sqrt{n^2 + 1}$ for all integers (or even that of $\frac{1}{2} + \sqrt{n^2 + 1}$ for half-integers) n; for example:

$$\sqrt{101} = 10 + \cfrac{1}{20 + \cfrac{1}{20 + \cfrac{1}{20 + \cfrac{1}{20 + \cdots}}}}.$$

Example 3.29. To give a nonexample to Abel's Theorem 3.1, let us examine the continued fraction expansion of the square root of the polynomial $D(t) = t^6 + 2t^3 + t + 1$. The Laurent series of $\sqrt{D(t)}$ around $t = \infty$ (same procedure as in the previous example) reads

$$t^3 + 1 + \frac{t^{-2}}{2} - \frac{t^{-5}}{2} - \frac{t^{-7}}{8} + \frac{t^{-8}}{2} + \frac{3t^{-10}}{8} - \frac{t^{-11}}{2} + \frac{t^{-12}}{16} - \frac{3t^{-13}}{4} + \frac{t^{-14}}{2}$$
$$- \frac{5t^{-15}}{16} + \frac{5t^{-16}}{4} - \frac{69t^{-17}}{128} + \cdots$$

and we calculate $a_0, a_1, \ldots$ just as before. Here is the list $a_0, a_1, \ldots, a_{13}$ provided by Merkert:

$$a_0 = t^3 + 1$$
$$a_1 = 2t^2$$
$$a_2 = \tfrac{1}{2}t$$
$$a_3 = -8t$$
$$a_4 = \tfrac{-1}{2}t + 2$$
$$a_5 = \tfrac{-1}{8}t - \tfrac{65}{128}$$

$$a_6 = -2048t - 8064$$

$$a_7 = \frac{-1}{65536}t + \frac{3}{32768}$$

$$a_8 = \frac{524288}{33}t + \frac{35651584}{1089}$$

$$a_9 = \frac{35937}{4259840}t - \frac{4886343}{138444800}$$

$$a_{10} = \frac{562432000}{81828549}t + \frac{52597667200}{1882056627}$$

$$a_{11} = \frac{-129861907263}{204068345000}t - \frac{161124749894097}{4665818640080000}$$

$$a_{12} = \frac{52089490911518125}{8659797998530734}t + \frac{7401227721243151250}{18830730747805081083}$$

$$a_{13} = \frac{72795420464181597893304}{219213673999487434840625}t - \frac{435427467400545923209648896}{645584269928490495605640625}.$$

This suggests that a_n is of degree 1, but that the height of the coefficients of a_n tends to $+\infty$ as $n \to \infty$. In particular, the sequence of polynomials $a_1, a_0, a_2, \ldots$ is not periodic. How to show directly that the Pell equation

$$x(t)^2 - (t^6 + 2t^3 + t + 1)y(t)^2 = 1$$

has no nontrivial solution? See Exercise 3.44, 2, below.

Although the periodicity of the continued fraction for $\sqrt{D(t)}$ is a very "rare" phenomenon, some periodicity survives in full generality. Indeed, we have the following:

Theorem 3.30 ([45, Theorem 1.1]). *Let k be an algebraically closed field of characteristic 0. Let $D \in k[t]$ be a polynomial of even degree, and let*

$$\sqrt{D(t)} = [a_0; a_1, a_2, a_3, \ldots]$$

be its continued fraction expansion. The sequence $\deg(a_0), \deg(a_1), \deg(a_2), \deg(a_3), \ldots$ is eventually periodic.

This analogue of Lagrange's theorem for the polynomial case seems not to have been noticed until now, as the most common behavior, which can be seen in many examples, is that all the degrees are equal to 1 or eventually constant. In particular, when $d \leq 3$ (or when the genus of the curve given by $u^2 = D(t)$ is 0), it may be seen that $\deg a_n$ is eventually constant in the non-Pellian case. More specifically, one can prove the following:

Proposition 3.31. *If $d \leq 3$ or the geometric genus is 0 (even if D is non-squarefree), either $D(t)$ is Pellian or there are only finitely many partial quotients with $\deg a_n > 1$.*

A proof of this, in the special case $D(t) = t^2(t^2 - 1)$, can be found in [45, Exa. 4.2]. We also point out that, if $d \geq 4$, this is not true anymore, as the following example (found by Merkert, see [22]) shows.

Example 3.32. The polynomial $D(t) = t^8 - t^7 - (3/4)t^6 + (7/2)t^5 - (21/4)t^4 + (7/2)t^3 - (3/4)t^2 - t + 1$ yields infinitely many partial quotients of degrees 1 and 2, with the periodic pattern of degrees $4, 1, 1, 2, 1, 1, 1, 1, 1, 1, 1, 1, 1, 2, 1, 1, 1, 1, 1, 1, 1, 1, 2, 1, \ldots$.

— **3.33.** We shall prove Theorem 3.30 in Section 3.9. It is not clear for which algebraic functions the sequence of degrees of partial quotients is periodic. The phenomenon seems not to be limited to square roots, for example, the convergents of the continued fraction expansion at infinity of $\sqrt[4]{t^4 + 3}$ are

$$a_0 = t$$

$a_1 = \frac{4}{3}t^3$	$a_{11} = \frac{9196}{1989}t^3$	$a_{21} = \frac{23896908}{3739405}t^3$	$a_{31} = \frac{115963743148}{14934083745}t^3$
$a_2 = \frac{2}{3}t$	$a_{12} = \frac{1326}{4807}t$	$a_{22} = \frac{7478810}{36698823}t$	$a_{32} = \frac{1422293690}{8416723293}t$
$a_3 = \frac{12}{5}t^3$	$a_{13} = \frac{19228}{3825}t^3$	$a_{23} = \frac{375143524}{56091075}t^3$	$a_{33} = \frac{370335824892}{46224544925}t^3$
$a_4 = \frac{10}{21}t$	$a_{14} = \frac{11050}{43263}t$	$a_{24} = \frac{37394050}{191649409}t$	$a_{34} = \frac{92449089850}{563920460631}t$
$a_5 = \frac{28}{9}t^3$	$a_{15} = \frac{173052}{32045}t^3$	$a_{25} = \frac{109513948}{15705501}t^3$	$a_{35} = \frac{76279096124}{9244908985}t^3$
$a_6 = \frac{30}{77}t$	$a_{16} = \frac{320450}{1341153}t$	$a_{26} = \frac{15397550}{82135461}t$	$a_{36} = \frac{92449089850}{580265981229}t$
$a_7 = \frac{2156}{585}t^3$	$a_{17} = \frac{54188}{9425}t^3$	$a_{27} = \frac{2956876596}{408035075}t^3$	$a_{37} = \frac{773687974972}{91199777825}t^3$
$a_8 = \frac{26}{77}t$	$a_{18} = \frac{64090}{284487}t$	$a_{28} = \frac{163214030}{903490071}t$	$a_{38} = \frac{269951342362}{1740797943687}t$
$a_9 = \frac{924}{221}t^3$	$a_{19} = \frac{7207004}{1185665}t^3$	$a_{29} = \frac{63402812}{8442105}t^3$	$a_{39} = \frac{633017434068}{72679207559}t^3$
$a_{10} = \frac{442}{1463}t$	$a_{20} = \frac{182410}{853461}t$	$a_{30} = \frac{163214030}{935191477}t$	$a_{40} = \frac{1889659396534}{12502094322843}t$

and their degrees clearly show a periodic pattern.

— **3.34.** As in the arithmetic situation, the solutions of the polynomial Pell equation form a group. We can identify it with a subgroup of the multiplicative group of the field $k(t)[u]/\langle u^2 - D \rangle$ by associating $(x, y) \mapsto x + yu$. It can be shown that the group of solutions of the Pell equation is isomorphic to $\mathbb{Z}/2\mathbb{Z}$ in the non-Pellian case, and to $\mathbb{Z}/2\mathbb{Z} \oplus \mathbb{Z}$ in the Pellian case. To check this, show that all solutions are generated by a solution of minimal degree.

— **3.35.** Let D be a nonsquare polynomial of even degree $2d$, and set $\sqrt{D} = [a_0; a_1, a_2, \ldots]$. It can be shown that $1 \leq \deg a_n \leq d$ holds for all n. The upper bound $\deg a_n = d$ holds for some $n > 0$ if and only if D is Pellian; if this is the case, then such values of n form an arithmetic progression. On the other hand, if D is squarefree and not Pellian, then the tighter upper bound $\deg a_n \leq d/2$ holds for all n big enough (see [45], Thm. 1.3 and the paragraph above it for details and a more precise statement).

— **3.36.** Another interesting line of inquiry concerns the heights of $a_n(t)$, $p_n(t)$, $q_n(t)$ over $\overline{\mathbb{Q}}$. The height of a nonzero polynomial $f \in \overline{\mathbb{Q}}[t]$, denoted $h(f)$, is the usual projective absolute (logarithmic) height of the vector of the coefficients.

The affine height of f is the affine height of the same vector; it is denoted by $h_a(f)$. It can be shown that when D is not Pellian, then the heights of the q_n grow quadratically in terms of n: $h(q_n) \gg n^2$; this follows from a more general theorem of Bombieri-Cohen [3], but can also be proved directly. For the partial quotients the following theorem holds.

Theorem 3.37 ([45, Thm. 1.5]). *Suppose that $D(t) \in \overline{\mathbb{Q}}[t]$ is squarefree and non-Pellian. Then $h(a_n) \ll n^2$. Also, there exists an integer $M = M_D$ such that*

$$\max\{h_a(a_{n-s}) \mid 0 \leq s \leq M\} \gg n^2$$

holds for large n.

We remark that this theorem cannot be recovered easily from the bounds on q_n and the recurrence relation satisfied by the q_n and the a_n and requires an independent proof.

— **3.38.** A question of McMullen [21] asks whether every real quadratic field $\mathbb{Q}(\sqrt{d})$ contains infinitely many periodic continued fractions $x = [\overline{a_0, a_1, \ldots}]$ such that $a_i \in \{1, 2\}$ for all $i = 1, 2, \ldots$. In her PhD thesis [18], Malagoli proved a function field analogue of this question:

Theorem 3.39 ([18, Theorem 5]). *Let k be a number field; then, for every nonsquare polynomial $D \in k[t]$ of even degree, not a square and with leading coefficient which is a square in k, there exists a polynomial $f \in k[t]$ such that the partial quotients of $f\sqrt{D}$ (except possibly for finitely many of them) have degree $= 1$.*

The proof of this theorem relies on the study of zeroes of the denominators $q_n(t)$ of the partial quotients, which appear infinitely often. This is of interest if we want to specialize t to an element of $\overline{\mathbb{Q}}$. In this context, Zannier proved the following result:

Theorem 3.40 ([45, Theorem 1.7]). *Let k be a number field and let $D \in k[t]$ be a polynomial of even degree. Then, for each $l \in \mathbb{R}$ there are only finitely many $\theta \in \overline{\mathbb{Q}}$ of degree $\leq l$ over k which are common zeroes of infinitely many $q_n(t)$.*

We point out that the proof of Malagoli's theorem deals also with the case of nonsquarefree $D(t)$ which complicates (also conceptually) the proofs.

— **3.41.** Another question which can be investigated regards how prime factors arise in denominators of polynomial continued fractions. This is strongly related to the problem of reducing polynomial continued fractions modulo a prime. In his thesis [22], Merkert studied this problem for the continued fractions of square roots of polynomials with rational coefficients in the nontrivial case that the continued fraction is not periodic (i.e., $D(t)$ is not Pellian). More precisely, he proved the following result:

Theorem 3.42 (Theorem 1, [22]). *If $D(t)$ is not Pellian, then for all primes p except finitely many, p appears in infinitely many polynomials a_n in a denominator of the coefficients.*

Notice that the primes excluded by the theorem are exactly the prime 2, any prime appearing already in the denominators of the coefficients of D, and those such that D_p (the reduction modulo p of D) is a square. The proof of this result is based on the comparison between the continued fractions of $\sqrt{D}$ and $\sqrt{D_p}$. This question was already studied in a series of papers [38], [39], [40] by van der Poorten, which analyzed whether the reduction of the convergents of $\sqrt{D}$ gives the convergents of $\sqrt{D_p}$ giving a theorem whose proof seems incomplete. In his thesis, Merkert completes the proof of van der Porten's theorem to prove his result (see [22, Thm. 7.2]). We also point out that these questions are related to the problem of reducing minimal solutions of the polynomial Pell equation, and has recently been used by Platonov [25] to construct hyperelliptic curves over $\mathbb{Q}$ of genus 2, where the Jacobian contains a torsion point of a specific order. These examples are relevant for the uniform boundedness conjecture for torsion points of abelian varieties.

— **3.43.** We present here to the interested reader three exercises about Pellian polynomials. The solutions are collected in Section 3.10.

Exercise 3.44. 1. Show that if $D \in \mathbb{Z}[t]$ is monic and irreducible over any quadratic extension of $\mathbb{Q}$, then D is not Pellian.
   2. Show that if $D \in \mathbb{Z}[t]$ is a monic polynomial, irreducible over $\mathbb{Q}$, and, for every prime p, not a square modulo p, then D is not Pellian.
   3. Show that if $D \in \mathbb{Z}[t]$ is monic, irreducible over $\mathbb{Q}_2(\sqrt{5})$, and not a square modulo 2, then D is not Pellian.

3.6   DISTRIBUTION OF PELLIAN POLYNOMIALS

In this section we give a criterion for the solvability of the polynomial Pell equation $x(t)^2 - Dy(t)^2 = 1$ in terms of a special point on the Jacobian of the hyperelliptic curve $u^2 = D(t)$, in the case where D is squarefree. This will allow us to study the solvability of the Pell equation for families of polynomials, and to connect the problem to the topic of unlikely intersections.

— **3.45.** Let k be an algebraically closed field of characteristic 0. Let $D(t) \in k[t]$ be a squarefree polynomial of even degree $\deg D = 2d > 0$. As D is squarefree, the affine curve given by the equation

$$u^2 = D(t)$$

is smooth. We denote by $C \in \mathbb{P}^2_k$ the corresponding smooth projective curve. We may cover C by two affine charts, one given by the affine curve above, and the other by the affine curve $v^2 = s^{2d}D(s^{-1})$, the gluing map between charts given by $(v,s) = (ut^{-d}, t^{-1})$ whenever it is defined. The genus of C is $g = d - 1 > 0$. Such a curve is called a *hyperelliptic curve*, the elliptic curves being those where $d = 2$. We may view C as a $2:1$ cover of $\mathbb{P}^1$ ramified at the $2d$ distinct zeroes of D. In particular $C \to \mathbb{P}^1$ is unramified at infinity. The projective curve C has thus two distinct points at infinity, corresponding to the two distinct roots of $s^{2d}D(s^{-1})$ around $s = 0$. We denote[4] these two points by ∞_+ and ∞_-. Let J be the Jacobian variety of C. We embed C into J via

$$j : x \mapsto \text{class of the divisor } (x) - (\infty_+) =: [(x) - (\infty_+)]$$

and write

$$\delta := j(\infty_-) = [(\infty_-) - (\infty_+)].$$

Notice that if the ground field k is not algebraically closed, then the points ∞_+ and ∞_- might not be defined over k. In this case they are conjugate points of degree two over k. With these notations, the following holds:

Theorem 3.46. *Let $D(t) \in k[t]$ be a polynomial of even degree and nonzero discriminant. With the notation from above, the polynomial D is Pellian if and only if $\delta \in J(k)$ is a torsion point.*

Proof. Suppose first that D is Pellian, so there exist polynomials $x(t)$ and $y(t) \neq 0$ satisfying $x^2 - Dy^2 = 1$. The nonconstant rational functions

$$\varphi_+ = x(t) + y(t)u \qquad \text{and} \qquad \varphi_- = x(t) - y(t)u$$

on C are regular on the affine part of C, so their divisors of poles are supported on $\{\infty_+, \infty_-\}$. Since $\varphi_+ \cdot \varphi_- = 1$, also their divisors of zeroes are supported at $\{\infty_+, \infty_-\}$, so we have

$$\text{div}(\varphi_+) = a(\infty_+) + b(\infty_-)$$

for integers a, b which are not both zero. The degree of $\text{div}(\varphi_+)$ is zero, hence $b = -a$ and thus $a\delta = \text{div}(\varphi_+)$. This shows that δ is a torsion point of order dividing a. Conversely, suppose that δ is torsion, so $a\delta$ is a principal divisor for some $a \neq 0$. Set $a\delta = \text{div}(\psi)$. We may write ψ as $x(t) + y(t)u$ on the affine part of C, where x and y are polynomials in t. The function $(x + yu)(x - yu) = x^2 - Dy^2$ is then a rational function on $\mathbb{P}^1$ whose divisor is supported at infinity, hence must be constant. Scaling x and y by a square root of this constant yields a solution of the Pell equation. $\qquad\square$

— **3.47.** Let k be an algebraically closed field. A polynomial $D(t) \in k[t]$ is Pellian if and only if the polynomial $cD(at + b)$ is Pellian for some $a, c \in k^*$ and

---

[4]After fixing the equation $u^2 = D(t) = d_0 t^{2d} + d_1 t^{2d-1} + \cdots$ it is possible to identify the two points by stipulating that $u \pm d_0^{1/2} t^d$ has a zero at $\infty_\pm$.

$b \in k$. A suitable substitution will bring a general polynomial $D(t)$ of even degree $2d$ into the form

$$D(t) = t^{2d} + t^m + a_1 t^{m-1} + \cdots + a_m$$

for some $m \leq 2d - 2$. We may consider the affine spaces $\mathbb{A}_k^m$ for $0 \leq m \leq 2d - 2$ as moduli for polynomials of degree $2d$ up to substitutions $D(t) \rightsquigarrow cD(at + b)$. As such, $\mathbb{A}_k^m$ contains a nonempty open subvariety $U \subseteq \mathbb{A}_k^m$ where the discriminant

$$\mathrm{disc}(t^{2d} + t^m + a_1 t^{m-1} + \cdots + a_m)$$

as a polynomial in $(a_1, \ldots, a_m)$ is nonzero. Over this open subvariety, the curves $u^2 = D(t)$ are smooth, and their Jacobians define a principally polarized abelian scheme J over U. On the boundary of U, the abelian scheme J degenerates. The abelian scheme J comes equipped with a section $\sigma : U \to J$ given by the divisor of points at infinity $(\infty_-) - (\infty_+)$. We may regard σ as a group homomorphism $\mathbb{Z} \to J$, hence as a peculiar 1-motive $M = [\mathbb{Z} \to J]$ over U. We want to understand the set

$$\{\lambda \in U \mid \sigma_\lambda \text{ is torsion in } J_\lambda\}, \tag{3.8}$$

that is, the set of those $\lambda \in U$ for which the 1-motive M_λ splits up to isogeny.

— **3.48.** Let us take an analytic viewpoint on the exceptional set (3.8). Let U be a simply connected complex manifold, and let $A \to U$ be a holomorphic family of complex tori of dimension g on U. We obtain a vector bundle $\mathrm{Lie}(A)$ of rank g over U. The kernel of the exponential map $\mathrm{Lie}(A) \to A$ is a local system of free $\mathbb{Z}$-modules of rank $2g$, which we may identify with the homology $H_1(A/U)$. Let $\omega_1, \ldots, \omega_{2g}$ be a basis of sections of this local system. We now may describe sections $\sigma : U \to A$ as functions $\beta : U \to \mathbb{R}^{2g}$ via the following correspondence.

$$\beta : U \to \mathbb{R}^{2g} \qquad \sigma(u) = \exp \sum_{i=1}^{2g} \beta_i(u) \omega_i(u)$$

We refer to β as *Betti map*. Notice that $\sigma(u_0)$ is a torsion point in the fibre A_{u_0} if and only if all coordinates of $\beta(u_0)$ are rational.

Let us consider the situation of 3.47, taking for simplicity as U a simply connected open subset of $\mathbb{C}^m$ where the discriminant $\mathrm{disc}(t^{2d} + a_1 t^{m-1} + \cdots + a_m)$ is nonzero. In this case, $2g = 2d - 2$ so the scheme J over U (given by the Jacobians) has relative dimension $d - 1$. The rank of the Betti map is defined as the rank of the Jacobian matrix of these Betti coordinates, at a certain point of U, with respect to any choice of real-coordinates x_j, y_j, where we can assume, for example, $z_j = x_j + iy_j$ are holomorphic coordinates on U, and x_j and y_j are the corresponding real and imaginary parts. For a general U the Betti map may be defined passing to the universal cover (as done in [1]). We call generic rank the maximal rank of this differential on S. The set where the rank decreases is a (proper) closed real-analytic subvariety; hence the set where the rank is the generic one is open and dense (since U is simply connected). Let u_0 be a point of U where the rank r is maximal (i.e., $=d-1$). By the implicit function theorem, the fiber $\beta^{-1}(\beta(u))$ is, in a neighbourhood of u_0, a real-analytic variety of dimension $2d-r$.

The rank of the Betti map has been intensively studied in [1] (see also [8, Sect. 1.2] for some general proofs in the case $d \leq 2$). In this case, the expectation is that the Betti map $\beta : U \to \mathbb{R}^{2d-2}$ has full rank almost everywhere, so we expect the fibers of β to be of complex dimension $d-1$. In particular, $\beta^{-1}(\mathbb{Q}^{2d-2})$ is a countable union of subvarieties of complex dimension $m-d-1$ (empty if $m < d-1$) in the ambient space U which has dimension m.

Suppose now that we are given a one-parameter family $D_\lambda(t)$ describing a curve L in U. Solely for dimension reasons, we expect $L \cap \beta^{-1}(x) = \varnothing$ for general $x \in \mathbb{R}^{2d-2}$. According to the philosophy of *unlikely intersections*, it is reasonable to expect that

$$L \cap \beta^{-1}(\mathbb{Q}^{2d-2}) = \{\lambda \in L \mid D_\lambda(t) \text{ is Pellian}\}$$

is a finite set, unless L has a very special shape. Indeed this has been proven in full generality by Masser and Zannier (see [23] for the special family $D_\lambda(t) = t^6 + t + \lambda$ and [new] for the general case).

Example 3.49. In the case $d=1$, the set U is a single point corresponding to the polynomial $t^2 - 1$, which is Pellian. The case $d=1$ becomes interesting if we add an arithmetic constraint and ask for the integers $n \neq 0$ such that the Pell equation

$$x(t)^2 - (t^2 + n)y(t)^2 = 1$$

has a nontrivial solution with $x(t), y(t) \in \mathbb{Z}[t]$. The answer was given by Nathanson in [24]: there is a nontrivial solution if and only if $n \in \{-2, -1, 1, 2\}$, and we can moreover describe all solutions.

Example 3.50. Consider the family of polynomials $D_\lambda(t) = t^4 + t + \lambda$. With the notation of 3.47, we are in the case $d=2$ and $m=1$. The discriminant of $D_\lambda(t)$ is $2^8\lambda^3 - 3^3$, so we will take for U the complex plane minus the three points $\frac{3}{8}\sqrt[3]{2}e^{2\pi i p/3}$ for $p=0,1,2$. One can show that, in this example, the Betti map $\mathbb{C} \supseteq U \to \mathbb{R}^2$ is locally surjective, so we expect countably many $\lambda \in U$ for which $D_\lambda(t)$ is Pellian. As a consequence of a theorem of Silverman-Tate [34], algebraic points $\lambda \in U$ for which $D_\lambda(t)$ is Pellian have bounded height. In particular, given any number field k, there are only finitely many $\lambda \in k$ for which $t^4 + t + \lambda$ is Pellian.

Example 3.51. Consider the family of polynomials $D_\lambda(t) = t^6 + t + \lambda$. With the notation of 3.47, we are in the case $d=3$ and $m=1$. The discriminant is $5^5 - 6^6\lambda^5$. In this case the Betti map $\mathbb{C} \supseteq U \to \mathbb{R}^4$ cannot be surjective, so it is unlikely that $\beta(u)$ has only rational coordinates. Let us compute a few terms of the continued fraction expansion of the square root of the polynomial $D_\lambda(t)$.

We may think for the moment that the field of coefficients is $\mathbb{Q}(\lambda)$. The Laurent series expansion of $\sqrt{D}$ at infinity reads

$$\sqrt{D(t^{-1})} = t^3 + \frac{t^{-2}}{2} + \frac{\lambda t^{-1}}{2} - \frac{t^{-7}}{8} - \frac{\lambda t^{-8}}{4} - \frac{\lambda^2 t^{-9}}{8} + \frac{t^{-12}}{16} + \frac{3\lambda t^{-13}}{16} + \frac{3\lambda^2 t^{-14}}{16}$$
$$+ \frac{\lambda^3 t^{-15}}{16} - \frac{5t^{-17}}{128} + \cdots$$

and has polynomial part t^3. We find

$$a_0 = t^3$$
$$a_1 = 2t^2 - 2\lambda t + 2\lambda^2$$
$$a_2 = -\frac{t}{2\lambda^3} - \frac{1}{2\lambda^2}$$
$$a_3 = -8\lambda^6 t + 16\lambda^7$$
$$a_4 = -\frac{t}{24\lambda^8 - 2\lambda^3} - \frac{16\lambda^5 - 1}{288\lambda^{12} - 48\lambda^7 + 2\lambda^2}$$
$$a_5 = -\frac{(1 - 12\lambda^5)^3 t}{8\lambda^9} - \frac{18432\lambda^{25} + 15360\lambda^{20} - 6400\lambda^{15} + 848\lambda^{10} - 48\lambda^5 + 1}{128\lambda^{18}}$$

and the expressions keep growing. According to Abel's theorem, $D_{\lambda_0}(t)$ is Pellian if and only if the specialized sequence of the a_i is periodic. Already from these few terms this periodicity seems unlikely. Indeed, it has been shown by Masser and Zannier that there are only finitely many $\lambda_0 \in U$ for which $D_{\lambda_0}(t)$ is Pellian (see [23]).

3.7   THE PELL EQUATION IN THE NONSQUAREFREE CASE

In this section, we analyze some examples of polynomial Pell equations with nonsquarefree D. These can be interesting for certain applications, and involve the study of so-called *generalized Jacobians*. Consider, for example, the family

$$D_\lambda(t) = t^2(t^4 + t^2 + \lambda t)$$

where λ varies over complex numbers such that $\mathrm{disc}(t^4 + t^2 + \lambda t) \neq 0$. The corresponding curves

$$u^2 = D_\lambda(t)$$

have a cusp at $(u, t) = (0, 0)$. The criterion in Theorem 3.46 is still valid when instead of the Jacobian of a smooth curve we consider the generalized Jacobian of a possibly singular projective curve (see Theorem 3.54 below). The theory of generalized Jacobians goes back to Rosenlicht [27], and a standard reference is

Chapter V in Serre's *Groupes algébriques et corps de classes*, [32]. For the curve above, the generalized Jacobian is an algebraic group G_λ for which the short exact sequence

$$0 \to \mathbb{G}_a \to G_\lambda \to E_\lambda \to 0$$

holds. It turns out that the extension is nonsplit (for a proof, see [33, p. 188] or [7, p. 249]). We can then recover a finiteness result analogously to the case of squarefree D; this has been done by H. Schmidt, in his PhD thesis [31], using again the Betti maps and involving in this case the Weierstrass $\wp$ and ζ functions.

— **3.52.** Let us give a short résumé on generalized Jacobians. Let C be a smooth projective curve of genus $g \geq 0$ over a field k, and let

$$\mathfrak{m} = \sum_{i=1}^{d} n_i P_i \tag{3.9}$$

be an effective divisor on C. We suppose that in (3.9) the P_i are distinct, so that d is the degree of the reduced divisor underlying $\mathfrak{m}$. We call $\mathfrak{m}$ a *modulus*. For a given rational function f on C, write $f \equiv 1 \bmod \mathfrak{m}$ if $\mathrm{ord}_{P_i}(1 - f) \geq n_i$ holds for each i. Given two divisors D and D' on C whose support is disjoint from the support of $\mathfrak{m}$, we say that D and D' are $\mathfrak{m}$-*equivalent* and write

$$D \sim_{\mathfrak{m}} D'$$

if there exists a rational function f such that $D - D' = \mathrm{div}(f)$ and $f \equiv 1 \bmod \mathfrak{m}$. The zealous reader may check that $\sim_{\mathfrak{m}}$ is indeed an equivalence relation. Set

$$\mathrm{Pic}^0_{\mathfrak{m}}(C) := \frac{\text{Divisors on } C \text{ of degree } 0 \text{ and support disjoint from } \mathfrak{m}}{\text{Divisors } \mathrm{div}(f) \text{ with } f \equiv 1 \bmod \mathfrak{m}}.$$

A first theorem of Rosenlicht [33, Chap.V, Prop. 2 and Thm. 1(b)] states that the functor $k' \mapsto \mathrm{Pic}^0_{\mathfrak{m}}(C \times_k k')$ is representable by a commutative connected algebraic group $G_{\mathfrak{m}}$ over k. We call $G_{\mathfrak{m}}$ the generalized Jacobian of the pair $(C, \mathfrak{m})$. Its dimension is g if $\mathfrak{m} = 0$ and $g + \deg(\mathfrak{m}) - 1$ if $\mathfrak{m} \neq 0$. If $\mathfrak{m} = 0$, we recover the usual Jacobian of C. If $\mathfrak{m}'$ divides $\mathfrak{m}$, there is a canonical surjective morphism $G_{\mathfrak{m}} \to G_{\mathfrak{m}'}$. In particular, there is a canonical short exact sequence

$$0 \to L_{\mathfrak{m}} \to G_{\mathfrak{m}} \to A \to 0$$

where $A = G_0$ is the Jacobian of C. A second theorem of Rosenlicht [33, V.13–V.17] concerns the structure of $L_{\mathfrak{m}}$. It states that $L_{\mathfrak{m}}$ is an affine algebraic group, isomorphic to the product of a torus T of dimension $d - 1$ (and gives its precise structure), and an additive group of dimension $\deg(\mathfrak{m}) - d$. As in (3.9), d is the degree of the reduced divisor underlying $\mathfrak{m}$, hence if $\mathfrak{m}$ is already reduced, $G_{\mathfrak{m}}$ is a semi-abelian variety.

Definition 3.53. Let C be a proper, but not necessarily smooth, curve over a field k. We call *generalized Jacobian* of C the generalized Jacobian $G_{\mathfrak{m}}$ of the pair $(\widetilde{C}, \mathfrak{m})$ as introduced in 3.52, where $\widetilde{C}$ is the normalization of C and $\mathfrak{m}$ the exceptional divisor on $\widetilde{C}$.

Theorem 3.54. *Let $D(t) \in \mathbb{C}[t]$ be a polynomial of even degree, and denote by $C \subseteq \mathbb{P}^2$ the projective curve given by the equation $u^2 = D(t)$. Let G be the generalized Jacobian of C, and let $\delta := [(\infty_-) - (\infty_+)]$ be the divisor of points at infinity on C. The polynomial D is Pellian if and only if $\delta \in G(\mathbb{C})$ is a torsion point.*

Proof. The proof is essentially the same as that of Theorem 3.46, and is left as an exercise.   □

Example 3.55. Consider the polynomial Pell equation $x(t)^2 - D_\lambda(t)y(t)^2 = 1$, where D_λ is the following pencil of nonsquarefree polynomials

$$D_\lambda(t) = t^2(t^4 + t^2 + \lambda t)$$

where λ varies over the complex numbers such that $\mathrm{disc}(t^4 + t^2 + \lambda t) \neq 0$. As already mentioned, the affine curve given by the equation $u^2 = D_\lambda(t)$ is singular at ∞ and at 0, and its generalized Jacobian is a nonsplit extension

$$0 \to \mathbb{G}_a \to G \to E_\lambda \to 0$$

where E_λ is the elliptic curve given by $u^2 = \widetilde{D}_\lambda := t^4 + t^2 + \lambda t$. If (x_0, y_0) is a solution of the Pell equation $x^2 - D_\lambda y^2 = 1$, then (x_0, ty_0) is a solution of $x^2 - \widetilde{D}_\lambda y^2 = 1$. In this way, solutions of the Pell equation $x^2 - D_\lambda y^2 = 1$ are in one-to-one correspondence with solutions $(\widetilde{x}, \widetilde{y})$ of $x^2 - \widetilde{D}_\lambda y^2 = 1$ such that $t \mid \widetilde{y}$. From the viewpoint of Theorem 3.54, this reflects the evident fact that any torsion point of G maps to a torsion point on E_λ. A point $g \in G(\mathbb{C})$ is torsion if and only if it maps to a torsion point in $E_\lambda(\mathbb{C})$ and moreover satisfies a "linear" condition.

Example 3.56. Consider the family of polynomials $D_\lambda(t) = (t - \lambda)^2(t^2 - 1)$ for λ varying in $\mathbb{C} \setminus \{\pm 1\}$. In this case, the solvability of the associated Pell equation is related to the study of some special torsion points on $\mathbb{G}_m$. Let us consider the projective curve H defined by the equation $u^2 = t^2 - 1$: its normalization has genus zero, so its Jacobian is trivial. Consider the two points $\xi_\lambda^\pm = (\lambda, \pm\sqrt{\lambda^2 - 1})$ of H with first coordinate equal to λ. A divisor A of degree 0 on H is always principal, so $A = \mathrm{div}(f)$ for some function f on H; hence, we have an homomorphism from divisors on H to $\mathbb{G}_m$ sending $A \mapsto \frac{f(\xi_\lambda^+)}{f(\xi_\lambda^-)}$. This yields indeed an isomorphism from the generalized Jacobian of H to $\mathbb{G}_m$. The divisor $(\infty_-) - (\infty_+)$ is equal to $\mathrm{div}(z)$, where $z = t + u$. Here, as before, we denote by ∞_+ the pole of the function $t + u$ and by ∞_- the pole of $t - u$. The image

of $\mathrm{div}(z)$ under the described isomorphism is equal to $\frac{z(\xi_\lambda^+)}{z(\xi_\lambda^-)} = (\lambda + \sqrt{\lambda^2 - 1})^2$.
This means that the polynomial $D_\lambda(t)$ is Pellian if and only if $\lambda + \sqrt{\lambda^2 - 1}$ is a root of unity in $\mathbb{G}_m$. Hence there are countably infinitely many $\lambda \in \mathbb{C}$ such that the polynomial D_λ is Pellian.

Example 3.57. Consider the family of polynomials $D_\lambda(t) = (t - \lambda)^2 (t - \lambda - 1)^2$ $(t^2 - 1)$. We can generalize the construction of the previous example. This time, we consider the two pairs of points $\xi_\lambda^\pm = (\lambda, \pm\sqrt{\lambda^2 - 1})$ and $\xi_{\lambda+1}^\pm = (\lambda + 1, \pm\sqrt{\lambda^2 + 2\lambda})$, and obtain an isomorphism from the generalized Jacobian to $\mathbb{G}_m^2$ by sending $\mathrm{div}(f)$ to $\left(\frac{f(\xi_\lambda^+)}{f(\xi_\lambda^-)}, \frac{f(\xi_{\lambda+1}^+)}{f(\xi_{\lambda+1}^-)} \right)$. Arguing as in the previous case, we conclude that $D_\lambda(t)$ is Pellian if and only if $\lambda + \sqrt{\lambda^2 - 1}$ and $\lambda + 1 + \sqrt{\lambda^2 + 2\lambda}$ are both roots of unity. This is equivalent to studying the torsion points on a curve in $\mathbb{G}_m^2$, that in our case is the curve of equation $x + x^{-1} = 2 + y + y^{-1}$. In general, these questions are related to Manin-Mumford type questions for $\mathbb{G}_m$, already asked by Lang, and proved by Ihara, Serre, and Tate in the case of curves (see [14]) and then generalized by Laurent [15] and independently by Sarnak-Adams [30] to higher dimension. For a survey on these questions, see also Zannier's book [43].

Example 3.58. Consider the family of polynomials $D_\lambda(t) = (t - 1)^2 (t^4 + t + \lambda)$. In this case, the generalized Jacobian G is an extension by $\mathbb{G}_m$ of an elliptic curve E. This elliptic curve E is the Jacobian of the relative quartic with equation $u^2 = t^4 + t^2 + \lambda$, and the extension is non-split in general. Also in this case we can apply the Pellian criterion to the section s of G defined by the class of the relative divisor $(\infty_-) - (\infty_+)$ on the quartic, for the strict linear equivalence attached to the node of the sextic $u^2 = D_\lambda(t)$ at $t = 1$ (see [33]). This case was studied in [6]: applying the main theorem to the generalized Jacobian, we have again a result of finiteness.

Exercise 3.59. Prove that there are countably infinitely many $\lambda \in \mathbb{C}$ such that the Pell equation $x^2 - (t^4 + t^2 + \lambda t)y^2 = 1$ has a nontrivial solution, but that there are only finitely many λ for which there is a solution (x, y) where x is monic.

3.8   A SKOLEM-MAHLER-LECH THEOREM FOR ALGEBRAIC GROUPS

The key ingredient in the proof of Theorem 3.30 is a theorem on algebraic groups reminiscent of the classical Skolem-Mahler-Lech theorem. This theorem states that for a sequence of elements in a field of characteristic zero $u_1, u_2, \ldots$ which is generated by a linear recurrence relation, there exist an integer N and a subset $R \subseteq \mathbb{Z}/N\mathbb{Z}$ such that

$$u_n = 0 \iff (n \bmod N) \in R$$

holds, with finitely many exceptions. For sequences of rational numbers, this theorem is due to Skolem (1933). Subsequent generalizations are due to Mahler for the case of number fields (1935), and to Lech for general fields of characteristic zero (1953).

— **3.60.**   A subset $A \subseteq \mathbb{Z}$ is called a *full arithmetic progression* if there exist integers a and $b \neq 0$ such that $A = \{a + bn \mid n \in \mathbb{Z}\}$ holds. Subsets of $\mathbb{Z}$ which are the union of a finite set and finitely many full arithmetic progressions form the closed sets of a topology on $\mathbb{Z}$.

Theorem 3.61 (Skolem-Mahler-Lech).   *Let k be a field of characteristic zero. Let $c_1, \ldots, c_r$ and $u_1, \ldots, u_r$ be elements of k with $c_r \neq 0$, and recursively define $u_n \in k$ by*

$$u_n = c_1 u_{n-1} + \cdots + c_r u_{n-r}$$

for all $n \in \mathbb{Z}$. The set $\{n \in \mathbb{Z} \mid u_n = 0\}$ is the union of a finite set and finitely many full arithmetic progressions.

We shall recover this theorem as a corollary of Theorem 3.63 below.

Classical proofs of this theorem use p-adic methods in one way or another. The corresponding statement in characteristic $p > 0$ does not hold. The question was studied by Derksen (2005), but there are already counterexamples by Lech (1953).

Example 3.62.   The sequence $u_1, u_2, \ldots$ in $\mathbb{F}_p(t)$ defined by $u_1 = 0$, $u_2 = 2t$, $u_3 = 3t^3 + 3t^2$, and for $n \geq 4$

$$u_n = (2t + 2)u_{n-1} - (t^2 + 3t + 1)u_{n-2} + (t^2 + t)u_{n-3}$$

has the closed expression $u_n = (t + 1)^n - t^n - 1$. The set $\{n \mid u_n = 0\}$ is the set of all powers of p, which cannot be written as the union of a finite set and finitely many arithmetic progressions.

For the proof of Theorem 3.30 we will need the following result:

Theorem 3.63 ([45, Corollary 3.3]).   *Let k be a field of characteristic zero and let G be an algebraic group over k. Let $X \subseteq G$ be a closed subvariety of G, and let $g \in G(k)$ be a rational point. The set*

$$\{n \in \mathbb{Z} \mid g^n \in X(k)\}$$

is the union of a finite set and finitely many full arithmetic progressions.

Proof of Theorem 3.63.   The proof of the main statement consists in a series of reductions to particular cases, until we are in the situation where G is commutative, defined over a p-adic field, and g is sufficiently close to the identity that it lies in the image of the p-adic exponential map. The final argument is then an application of elementary p-adic analysis.

By replacing G by the Zariski closure of $\{g^n \mid n \geq 1\}$ we may, without loss of generality, assume that G is commutative and that $\{g^n \mid n \geq 1\}$ is Zariski dense in G. Let $G_0, \ldots, G_n$ be the connected components of G, where G_m is the component of g^m. The group G/G_0 is isomorphic to $\mathbb{Z}/n\mathbb{Z}$, generated by the class of $g \in G_1(k)$. If the statement of the theorem holds for the closed subvarieties $g^{-m}(X \cap G_m)$ of G_0 and the element $g^n \in G_0(k)$, then it holds for $X \subseteq G$ and $g \in G(k)$, hence we also may suppose that G is connected.

We are now in the situation where the group G is commutative and connected, and $\{g^n \mid n \geq 1\}$ is dense in G. If $X = G$, then the set $\{n \in \mathbb{N} \mid g^n \in X\}$ is all of $\mathbb{Z}$ and we are done. Suppose then that $X \neq G$, and let us show that the set $\{n \in \mathbb{N} \mid g^n \in X\}$ is indeed finite. In other words, we show that for any infinite subset $A \subseteq \mathbb{N}$ the set of points $\{g^a \mid a \in A\}$ is dense in G. Fix an infinite subset $A \subseteq \mathbb{N}$ and a rational function f on G such that $f(g^a) = 0$ for all $a \in A$. We must show that f is zero, and we will do so by using properties of p-adic analytic maps. In order to move to a p-adic setting, let us choose and still denote by G a model of G over $\mathrm{spec}(R)$ for some finitely generated integral ring R, such that the point $g \in G(k)$ extends to a point $g \in G(R)$. For some sufficiently big prime number p, there exist a finite extension K of $\mathbb{Q}_p$ and an embedding R into the ring $\mathcal{O}_K$ of integers of K. We may also assume that f has good reduction modulo p. It suffices to show that the set of points $\{g^a \mid a \in A\} \subseteq G(K)$ is dense in G, viewed as an algebraic group over K.

We are now in the situation where the group G is defined over a finite extension K of $\mathbb{Q}_p$ with a model over $\mathcal{O}_K$, and g is an integral point of G, that is, $g \in G(\mathcal{O}_K) \subseteq G(K)$. With its p-adic topology the group $G(K)$ is a topological group, and $G(\mathcal{O}_K) \subseteq G(K)$ is a compact open subgroup. There exists a p-adic analytic group homomorphism $e : \mathcal{O}_K^d \to G(\mathcal{O}_K)$ which is a homeomorphism of $\mathcal{O}_K^d$ onto its image (the map e is the p-adic exponential map; see [12] for an elementary treatment). The integer d is the dimension of G as an algebraic group. Since $G(\mathcal{O}_K)$ is compact we have $g^n \in e(\mathcal{O}_K)$ for some sufficiently big $n \geq 1$. Let us write $\xi = e^{-1}(g^n)$. By partitioning A into congruence classes modulo n and passing to one of these subsets, we may assume that all elements of A are pairwise congruent modulo n, and so we write $a = a'n + r$ with a fixed r for all elements $a \in A$.

Let us now consider the map $\phi : \mathcal{O}_K \to \mathcal{O}_K$ given by $\phi(z\xi) = f(g^{nz+r})$. For every $a \in A$ we have that a' is a zero of the locally analytic function $z \mapsto \phi(z\xi)$, but this is only possible if the function is identically zero, which implies that also f is identically zero as we wished to show. $\qquad \square$

We show now how the classical Skolem-Mahler-Lech theorem can be derived from the more general Theorem 3.63.

Proof of Theorem 3.61. Let $u_1, u_2, \ldots$ be a sequence of elements of k defined by its initial terms $u_1, \ldots, u_r \in k$ and a linear recurrence relation, say

$$u_n = c_1 u_{n-1} + \cdots + c_r u_{n-r}$$

for all $n > r$. In order to study the nature of the set $\{n \in \mathbb{N} \mid u_n = 0\}$ we may assume that k is algebraically closed, hence we may suppose that there exist $\alpha_1, \ldots, \alpha_r \in k$ and polynomials $P_1, \ldots, P_r \in k[t]$ such that

$$u_n = \sum_{i=1}^{r} P_i(n) \alpha_i^n$$

holds for all $n \geq 0$. We can consider the closed algebraic group $G := \mathbb{G}_a \times \mathbb{G}_m^r$ over k, the subvariety $X \subseteq G$ defined by

$$X = \{(y, z_1, \ldots, z_r) \mid P_1(y) z_1 + \cdots + P_r(y) z_r = 0\},$$

and the point $g = (1, \alpha_1, \ldots, \alpha_r) \in G(k)$. We have $g^n \in X(k) \iff u_n = 0$, hence Theorem 3.61 is indeed a consequence of Theorem 3.63.   □

3.9   PERIODICITY OF THE DEGREES OF THE PARTIAL QUOTIENTS

In this section we prove Theorem 3.30, which states that given a polynomial $D(t) \in K[t]$ where K is an algebraically closed field of characteristic zero, the sequence of degrees of the partial quotients in the continued fraction expansion of $\sqrt{D(t)}$ is periodic. If $D(t)$ is a square, the assertion holds trivially, so we will assume not to be in this case. Moreover, for simplicity of exposition we will only consider the case where D is squarefree, even if the reduction to this case is not immediate. The general case, which involves the use of generalized Jacobians, is treated in [45].

— **3.64.**   We call a sequence $(x_n)_{n \geq 0}$ *eventually periodic* if there exist integers $N \geq 1$ and $L \geq 1$ such that $x_{n+L} = x_n$ holds for all $n > N$. We call a subset $X \subseteq \mathbb{N}$ *eventually periodic* if its characteristic function, viewed as a sequence, is eventually periodic. In other words, a subset $X \subseteq \mathbb{N}$ is eventually periodic if, up to a finite set, it is a finite union of arithmetic progressions.

— **3.65.**   We fix once and for all a squarefree complex polynomial $D(t) \in K[t]$ of even degree $2d > 2$, and denote by

$$\sqrt{D(t)} = [a_0; a_1, a_2, a_3, \ldots]$$

the continued fraction expansion of $\sqrt{D}$. We denote by p_n/q_n the convergents, where p_n and q_n are the polynomials obtained from the recurrence relations (3.3). We set

$$l_n := \deg a_n$$

to ease the notation. Let us denote by $\mathcal{C}$ the non-singular model of the curve given by the equation $u^2 = D(t)$ as introduced in Section 3.6, and let J be the

Jacobian variety of C. The curve C has genus $g = d - 1$, so J is an abelian variety of dimension g. As done before, let us call ∞_+ and ∞_- the two points at infinity of C. The canonical embedding $j : C \to J$ sends $y \in C$ to the class of $(y) - (\infty_+)$. Define

$$\delta := j(\infty_-) = [(\infty_-) - (\infty_+)]$$

and, for $0 \leq m \leq g$, the closed, irreducible subvariety $W_m \subseteq J$ given by

$$W_m = \{j(y_1) + j(y_2) + \cdots + j(y_m) \mid y_1, \ldots, y_m \in C\},$$

which has dimension m. In particular $W_g = J$.

—— **3.66.** We will work with rational functions on the hyperelliptic curve C of the form $f = p - qu$, where p and q are polynomials in the variable t. Such a function is regular on the affine part of C, i.e., on the affine curve given by the equation $u^2 = D(t)$. An affine neighborhood of the points ∞_+ and ∞_- is given by the curve of equation

$$v^2 = s^{2d} D(s^{-1}) \tag{3.10}$$

as we have already seen in 3.45. Using the substitution rule $(u, t) = (vs^{-d}, s^{-1})$, the rational function $f(u, t) = p(t) - q(t)u$ transforms to $g(v, s) = p(s^{-1}) - q(s^{-1})$ vs^{-d}. Denoting by $c \in \mathbb{C}$ the leading coefficient of $D(t)$, the points ∞_+ and ∞_- correspond to the solutions $(v, s) = (\pm\sqrt{c}, 0)$ of (3.10). We can understand the behavior of f at the two points at infinity as follows: write $v = s^{-d}\sqrt{D(s^{-1})}$ around $s = 0$ as a Laurent series

$$v = s^{-d}\sqrt{D(s^{-1})} = \pm \sum_{n \geq -d} c_n s^n.$$

The behavior of f at the two points at infinity is then the one of the series

$$p(s) \pm q(s)v = p(s) \pm q(s) \sum_{n \geq -d} c_n s^n$$

around $s = 0$, the sign depending on the choice of the signs for ∞_+ and ∞_-. In particular, if p_n/q_n is a convergent in the continued fraction expansion of $\sqrt{D}$, then $p_n - q_n u$ has a pole of order $\deg p_n + d$ at one point, and a zero of order $\deg q_n + \deg a_n$ at the other point at infinity.

—— **3.67.** Our goal is to prove that the sequence $(\deg a_n)_n$ is eventually periodic. By (3.3), we have $\deg(p_{n+1}) = \deg(p_n) + \deg(a_n)$, hence to show that $(\deg a_n)_n$ is eventually periodic amounts to show that the set

$$B := \{\deg p_n \mid n \geq 1\}$$

is, up to a finite set, a union of finitely many arithmetic progressions. Let us introduce for $l \geq 1$ the following three sets:

$$A(l) := \{k \geq 1 \mid k \cdot \delta \in W_{d-l}\};$$
$$B(l) := \{k \geq 1 \mid \exists n \geq 1 \text{ with } \deg(p_n) = k, \ \deg(a_n) = l\};$$
$$C(l) := \{k \geq 1 \mid \exists n \geq 1, h \geq 1 \text{ with } \deg(p_n) = k - h, \ \deg(a_n) = l + h\}.$$

Notice that $A(d) = \varnothing$, $A(1) = \mathbb{N} \setminus \{0\}$, and, for every $i = 1, \ldots, d-1$ we have that $A(i+1) \subseteq A(i)$. Furthermore, as the sequence of the degrees of the p_n is strictly increasing, the sets $B(l)$ are all disjoint.

Theorem 3.63 states that for any integer $l \geq 0$, the set $A(l)$ is eventually periodic. We have

$$B = \bigcup_{l=1}^{d} B(l),$$

so it suffices to show for each l individually that $B(l)$ is eventually periodic. We will do this essentially by a "downward" induction on l, noting that $B(l)$ is empty for $l > d$. The bulk of the work consists of relating the sets $A(l)$, $B(l)$, and $C(l)$, which we will do in the following lemmas.

Lemma 3.68. *The inclusion $B(l) \subseteq A(l)$ holds for all $l \geq 1$.*

Proof. To show this, we consider the rational functions $\varphi_n := p_n - u q_n$ on the curve $\mathcal{C}$. After choosing signs suitably, φ_n has a zero at ∞_+ of order $\deg q_n + l_n$, and a pole at ∞_- of order $\deg q_n + d$, which is in fact the only pole. We then have

$$\operatorname{div}(\varphi_n) = (\deg q_n + l_n)(\infty_+) - (\deg q_n + d)(\infty_-) + \sigma_n$$
$$= -(\deg p_n)((\infty_-) - (\infty_+)) - (d - l_n)(\infty_+) + \sigma_n,$$

where σ_n is an effective divisor of degree $d - l_n$ of the form

$$\sigma_n = \sum_{i=1}^{d-l_n} (x_i)$$

with $x_i \neq \infty_\pm$, and we have used that $\deg p_n = \deg q_n + d$. Considering the corresponding linear equivalence classes, we have that

$$(\deg p_n)\delta = [\sigma_n - (d - l_n)(\infty_+)] = \sum_{i=1}^{d-l_n} [(x_i) - (\infty_+)] = \sum_{i=1}^{d-l_n} j(x_i),$$

hence $(\deg p_n)\delta \in W_{d-l_n}$ as wanted. $\square$

On a side note, using the fact that $W_{d-l_n} = W_{g-l_n-1}$, we can observe that in some sense we "usually" have $l_n = \deg a_n = 1$, since otherwise we would have $(\deg p_n)\delta \in W_{g-1}$, and $W_{g-1} \subsetneq J$ is a subvariety of codimension 1.

We also remark that a similar argument combined with a more general version of Theorem 3.63 implies, for instance, that if the Jacobian is simple, then either $\deg a_n = 1$ for all large n or δ is a torsion point, i.e., we are in the Pellian case. Since a generic curve has a simple Jacobian, this justifies the assertion that *"usually, almost all the a_n have degree 1."*

Lemma 3.69.   *Let us consider the set $C := \bigcup_{i \geq 1} C(i)$. Then, for every $l \geq 1$, the sets $B(l)$ and C are disjoint.*

Proof.   We argue by contradiction. Let $k \geq 1$ be an element of C and also of $B(l)$. As C is the union of the $C(i)$ for all $i \geq 0$ (and $C(i) = \emptyset$ for all $i \geq d$), then there exists $1 \leq r \leq d-1$ such that $k \in C(r)$. Hence, there exist by definition integers $n, m, h \geq 1$ such that

$$\deg p_m = k - h \qquad \deg a_m = r + h \qquad \deg p_n = k \qquad \deg a_n = l$$

holds. Moreover, notice that $\deg q_n - \deg q_m = \deg p_n - \deg p_m = h$. Consider now the rational function

$$q_n p_m - p_n q_m = q_n(p_m - u q_m) - q_m(p_n - u q_n);$$

then, on C it has a zero at ∞_+ of order at least the minimum of $\deg q_m + \deg a_m - \deg q_n = r$ and $\deg q_n + \deg a_n - \deg q_m = l + h$. We conclude that $q_n p_m - p_n q_m$ has a zero of order at least 1 at ∞_+, hence is identically zero since it is a polynomial in t which we just considered as a rational function on C via the $2:1$ covering $C \to \mathbb{P}^1$. Therefore as p_n/q_n and p_m/q_m are convergents of the continued fraction of $\sqrt{D}$, the equality $q_n p_m = p_n q_m$ implies $m = n$ and $k - h = \deg(p_m) = \deg(p_n) = k$, contradicting the assumption that h is positive.   □

Lemma 3.70.   *For every $l \geq 1$, the intersection of $B(l)$ and $A(l+1)$ is at most finite; more precisely, if $k \in B(l) \cap A(l+1)$ then $k \leq \frac{d-l-1}{2}$.*

Proof.   We argue by contradiction. Let $k \geq 1$ be an element of $A(l+1)$ and also of $B(l)$. Since $k\delta \in W_{d-(l+1)}$, there exists an effective divisor σ of degree $d - (l+1)$ such that

$$k\delta = [\sigma - (d-l-1)(\infty_+)].$$

This implies that there exists a rational function φ over C such that

$$\operatorname{div}(\varphi) = -k((\infty_-) - (\infty_+)) + \sigma - (d-l-1)(\infty_+). \tag{3.11}$$

Since the divisor of poles of φ is supported only on $\{\infty_+, \infty_-\}$, then there exist two polynomials p and q in $K[t]$ such that $\varphi = p - qu$. Moreover, since

$k \in B(l)$, there exists an integer $n \geq 1$ such that $\deg p_n = k$ and $\deg a_n = l$. Now, the function $\varphi_n = p_n - q_n u$ vanishes at ∞_+ with order $\deg q_n + l = k - d + l$. If we consider the rational function $pq_n - qp_n = q_n \varphi - q \varphi_n$, it has an order at ∞_+ at least

$$\min\{-\deg q_n + \mathrm{ord}(\varphi), -\deg q + \mathrm{ord}\varphi_n\} \geq \min\{l+1, -\deg q + k - d + l\}. \tag{3.12}$$

We want now to estimate the degree of q. If we consider the function $\varphi' = p + qu$, it will have order $\geq -k$ at ∞_+ and order $\geq k - d + l + 1$ at ∞_-, since it is the composition of the standard involution $(t, u) \mapsto (t, -u)$ of $K(\mathcal{C})$ with φ. Writing $uq = \frac{1}{2}(\varphi + \varphi')$, we have that

$$\mathrm{ord}_{\infty_+}(uq) \geq \min\{k - d + l + 1, -k\}.$$

If $k > \frac{d-l-1}{2}$, then the previous minimum is exactly $-k$, and we obtain that $-\deg q \geq -k + d$. Using (3.12), we have that

$$\mathrm{ord}_{\infty_+}(\varphi) \geq \min\{l+1, l\} \geq 1.$$

This means that $qp_n - pq_n$ vanishes at infinity, so as before it has to be identically zero. Since we have $\deg q \leq k - d$ and p_n and q_n are coprime, this implies that, after replacing p and q by suitable scalar multiples, $p = p_n$ and $q = q_n$. But then the divisorial relation (3.11) shows that $\deg a_n \geq l+1$, which contradicts our assumption. Hence we proved that if $k \in B(l) \cap A(l+1)$, then $k \leq \frac{d-l-1}{2}$ as wanted.   □

To ease the notation, for every $l \geq 1$ we have $B(l) = D(l) \cup (A(l) \setminus (A(l+1) \cup C_{>l}))$.

Lemma 3.71. *For every $l \geq 1$, we have that $B(l) = D(l) \cup (A(l) \setminus (A(l+1) \cup C_{>l}))$.*

Proof. The inclusion $\subseteq$ was shown in Lemmas 3.68, 3.69, and 3.70, so it is enough to prove that the inclusion $\supseteq$ holds as well. By definition, $D(l) \subseteq B(l)$, so we have only to care about the second set. Take $k \in A(l) \setminus (A(l+1) \cup C_{>l})$; then, we have $k\delta \in W_{d-l}$; moreover, we can assume that $k \notin D(l)$. As in the proof of Lemma 3.70, there exist two polynomials p and q in $K[t]$ such that the divisorial relation

$$\mathrm{div}(p - qu) = -k((\infty_-) - (\infty_+)) + \sigma - (d - l)(\infty_+) \tag{3.13}$$

holds, where σ is an effective divisor of degree $d - l$. Moreover, as in the previous lemma, we have that if $k \notin D(l)$, then $\deg q \leq k - d$.

A priori p and q need not be coprime, but we can prove the quotient p/q is a convergent in the continued fraction expansion of $\sqrt{D}$. Indeed, if r is the greatest common divisor between p and q, we can set $p = rp'$ and $q = rq'$. It is now easy to see that the rational function $p' - uq'$ has a zero at ∞_+ of order at least $k - d + l \geq \deg(q') + 1$, hence $p'/q' = p/q$ is a convergent by Proposition 3.20. Eventually replacing r by a suitable scalar multiple, this implies that there exists $n \geq 1$ such that $p = rp_n$ and $q = rq_n$.

Before showing that p and q are actually coprime, let us prove that the support of σ does not contain ∞_+. Indeed if this is not the case, then we can write $\sigma = \sigma' + (\infty_+)$. Then we find

$$\operatorname{div}(p - qu) = -k(\infty_-) - (\infty_+) + \sigma' - (d - l - 1)(\infty_+),$$

which in turn implies that $k \in A(l+1)$ contrary to our assumption.

Let us finally show that p and q are coprime. As σ is not supported at ∞_+, we have that the order of $p - uq$ at ∞_+ is exactly $k - d + l$. On the other hand, the order of $p - uq$ at ∞_+ is also equal to $\operatorname{ord}_{\infty_+}(r(p_n - q_n u)) = -\deg r + \deg q_n + \deg a_n$. Using that $\deg q_n = \deg p_n - d$, we have that $\deg a_n + \deg p_n = k + l + \deg r$. Furthermore, $\deg p_n \leq k - \deg r$, hence $k \in C(l + \deg r)$ contradicting the hypothesis that $k \notin C_{>l}$ if $\deg r \geq 1$.

This implies that, eventually taking suitable scalar multiples, $p = p_n$ and $q = q_n$, hence $k - \deg p_n$ and $l = \deg a_n$, proving that $k \in B(l)$ as wanted. $\square$

We will finally prove Theorem 3.30 for squarefree D using an inductive argument on l.

Proof of Theorem 3.30 for squarefree D.   As we have noticed in 3.67, to prove that the sequence of $(\deg a_n)_n$ is eventually periodic it suffices to prove that for every $l \geq 1$ the set $B(l)$ is eventually periodic. This is certainly true for $l \geq d + 1$, because in this case $B(l)$ is empty.

We proceed by downward induction on l. Fix $l \geq 1$, and suppose that $B(l')$ is eventually periodic for all $l' > l$. For every $i > l$, we can write $C(i)$ as

$$C(i) = \bigcup_{h \geq 1} \{B(i+h) + h\};$$

hence, using the inductive hypothesis, any of these $C(i)$ for $i > l$ is eventually periodic. This implies that $C_{>l} = \bigcup_{i>l} C(i)$ is also eventually periodic. Finally, applying Theorem 3.63, we have that the sets $A(l)$ and $A(l+1)$ are eventually periodic. From Lemma 3.71 and the fact that $D(l)$ is at most finite, we deduce that $B(l)$ is eventually periodic as well, proving the theorem. $\square$

3.10   SOLUTIONS TO THE EXERCISES

Proposition 3.72.   *Let $D(t) \in \mathbb{Z}[t]$ be a monic polynomial with the property that D is irreducible over any quadratic extension of $\mathbb{Q}$. Then $D(t)$ is not Pellian.*

Proof.   Assume that D is Pellian, i.e., that there exist polynomials $A, B \in \mathbb{Q}[t]$ with $B \neq 0$ and
$$A^2 - B^2 D = 1.$$

Suppose moreover that this solution is minimal in terms of the degree of the polynomial A. Rearranging this equality and factorizing we obtain

$$(A+1)(A-1) = B^2 D.$$

The two polynomials on the left-hand side are coprime in $\mathbb{Q}[t]$, and D is irreducible, therefore we can write (up to changing the sign of A)

$$\begin{cases} A+1 = E^2/\alpha \\ A-1 = \alpha C^2 D \end{cases}$$

where $\alpha \in \mathbb{Q}^*$ and $C, E \in \mathbb{Q}[t]$ are two polynomials such that $B = CE$. By taking the difference of these two equations and multiplying by α we obtain

$$2\alpha = E^2 - \alpha^2 C^2 D$$

which leads to

$$D = \frac{E^2 - 2\alpha}{\alpha^2 C^2}.$$

Let now $\beta = \sqrt{2\alpha}$. Then we have the factorization over $\mathbb{Q}(\beta)$,

$$D = \frac{(E+\beta)(E-\beta)}{\alpha^2 C^2}.$$

Notice that $\beta \notin \mathbb{Q}$, otherwise $(E/\beta, \alpha C/\beta)$ would be a solution to the original Pell equation with $\deg E < \deg A$. Hence we see that on the right-hand side, after cancelling the denominator, there must be an even number of irreducible factors in $\mathbb{Q}(\beta)[t]$, against the hypothesis on D. $\square$

Proposition 3.73.   *Let $D \in \mathbb{Z}[t]$ be a monic polynomial irreducible over $\mathbb{Q}$, and assume that, for every prime p, D is not a square modulo p. Then, $D(t)$ is not Pellian.*

Proof.   Assume that D is Pellian, i.e., there exist polynomials $A, B \in \mathbb{Q}[t]$, with $B \neq 0$, such that
$$A^2 - B^2 D = 1. \tag{3.14}$$

Suppose moreover that this solution is minimal in terms of the degree of the polynomial A. If we get rid of denominators, we obtain

$$a^2 A^2 - b^2 B^2 D = u^2,$$

with $A, B \in \mathbb{Z}[t]$ primitive polynomials and $a, b, u \in \mathbb{Z}$ with a, b, u pairwise coprime.

Suppose that $u^2 \neq 1$; then, if p is a prime dividing u and we reduce mod p, we have $b^2 B^2 D \equiv a^2 A^2 \mod p$, hence D is a square modulo p, contradicting the hypothesis. Moreover, if $a^2 \neq 1$ and we reduce modulo a prime p dividing a, then we have $B^2 D \equiv (b^2)^{-1} \mod p$ that is impossible as D is monic by hypothesis. So we reduce to an equation of the form

$$A^2 - b^2 B^2 D = 1,$$

with $A, B \in \mathbb{Z}[t]$ primitive polynomials and $b \in \mathbb{Z}$.

Notice that b must be even, otherwise we would have that D is a square modulo 2 contradicting the hypothesis. Let us then write $b = 2^k b'$ with $k \geq 1$ and $(b', 2) = 1$.

Let us rewrite our equation as $A^2 - 1 = b^2 B^2 D$, i.e.,

$$(A+1)(A-1) = b^2 B^2 D.$$

Since D is irreducible, D will divide one of the two factors. As $(A+1, A-1) = 2$, then we can write (up to changing the sign of A)

$$\begin{cases} A + \epsilon = 2^\alpha e^2 E^2 \\ A - \epsilon = 2^{2k-\alpha} c^2 C^2 D \end{cases}$$

where $B = CE$ with $C, E \in \mathbb{Z}[t]$ and $(C, E) = 1$, $b^2 = 2^{2k} c^2 e^2$ with $(c, e) = 1$, $\epsilon = \pm 1$, and $\alpha = 1$ or $\alpha = 2k - 1$.

Taking the difference between these two equations we have

$$\pm 2 = 2^\alpha e^2 E^2 - 2^{2k-\alpha} c^2 C^2 D,$$

hence, dividing by 2, we have

$$\pm 1 = 2^{\alpha-1} e^2 E^2 - 2^{2k-\alpha-1} c^2 C^2 D.$$

Notice that we can exclude that $\alpha = 2k - 1$, because otherwise D would be a square modulo 2, contradicting the hypothesis. So $\alpha = 1$ and the equation reduces to

$$e^2 E^2 - 2^{2k-2} c^2 C^2 D = \pm 1.$$

As done before, we have that $e^2 = 1$, otherwise if p is a prime dividing e $(p \neq 2)$ and we reduce modulo p, we have $2^{2k-2}c^2C^2D \equiv \pm 1 \mod p$ which is impossible as D is monic.

So, we reduce to an equation of the form

$$E^2 - 2^{2k-2}c^2C^2D = \pm 1.$$

If the sign on the right-hand side is a plus, then $(E, 2^{k-1}cC)$ is again a solution of the Pell equation (3.14) for D, with $\deg E < \deg A$. But A was chosen to have minimal degree, which gives a contradiction.

Therefore, we can assume that

$$E^2 - 2^{2k-2}c^2C^2D = -1. \tag{3.15}$$

If $2k - 2 = 0$, then D is a square modulo 2, which contradicts our hypothesis.

On the other hand, if $2k - 2 \geq 2$, we have that $E^2 \equiv 1 \mod 2$, hence $E = 2F + 1$ with $F \in \mathbb{Z}[t]$. If we substitute in (3.15), we have

$$2^{2k-2}c^2C^2D - (1 + 2F)^2 = 1,$$

hence

$$2^{2k-2}c^2C^2D - 4F^2 - 4F = 2,$$

which gives a contradiction reducing modulo 4, thus concluding the proof. □

Remark 3.74. Notice that, in the previous proposition, the hypothesis *"for every prime p, D is not a square modulo p"* cannot be improved. Take for example

$$D(t) = t^2 + t + 1.$$

Then, D is Pellian, in fact:

$$\left(\frac{8}{3}t^2 + \frac{8}{3}t + \frac{5}{3}\right)^2 - \left(\frac{8}{3}t + \frac{4}{3}\right)^2 (t^2 + t + 1) = 1.$$

Moreover, D is monic and irreducible over $\mathbb{Q}$ and D is not a square modulo p for every prime $p \neq 3$ (we have instead that $t^2 + t + 1 \equiv (t + 2)^2 \mod 3$). Furthermore, notice that 3 is exactly the prime which appears in the denominators of the solution of the Pell equation, as also shown in the proof of the proposition.

Proposition 3.75. *Let $D \in \mathbb{Z}[t]$ be a monic polynomial. Assume D is irreducible over $\mathbb{Q}_2(\sqrt{5})$ and not a square modulo 2; then D is not Pellian.*

Proof. Assume that D is Pellian, i.e., there exist polynomials $A, B \in \mathbb{Q}[t]$, with $B \neq 0$, such that

$$A^2 - B^2D = 1. \tag{3.16}$$

Suppose moreover that this solution is minimal in terms of the degree of the polynomial A. If we get rid of denominators, we obtain

$$a^2 A^2 - b^2 B^2 D = u^2,$$

with $A, B \in \mathbb{Z}[t]$ primitive polynomials and $a, b, u \in \mathbb{Z}$ with a, b, u pairwise coprime.

Suppose $a^2 \neq 1$; if we reduce modulo a prime p dividing a, then we have $B^2 D \equiv (b^2)^{-1} \mod p$ that is impossible as D is monic by hypothesis. So we reduce to an equation of the form

$$A^2 - b^2 B^2 D = u^2,$$

with $A, B \in \mathbb{Z}[t]$ primitive polynomials and $b, u \in \mathbb{Z}$.

Notice that b must be even, otherwise we would have that D is a square modulo 2 contradicting the hypothesis. Let us then write $b = 2^k b'$ with $k \geq 1$ and $(b', 2) = 1$.

Let us rewrite our equation as $A^2 - u^2 = b^2 B^2 D$, i.e.

$$(A + u)(A - u) = b^2 B^2 D.$$

Since D is irreducible, D will divide one of the two factors. As $(A + u, A - u) = 2$, we can write (up to changing the sign of A and u)

$$\begin{cases} A + u = 2^\alpha e^2 E^2 \\ A - u = 2^{2k-\alpha} c^2 C^2 D \end{cases}$$

where $B = CE$ with $C, E \in \mathbb{Z}[t]$ and $(C, E) = 1$, $b^2 = 2^{2k} c^2 e^2$ with $(c, e) = 1$, and $\alpha = 1$ or $\alpha = 2k - 1$.

Taking the difference between these two equations we have

$$2u = 2^\alpha e^2 E^2 - 2^{2k-\alpha} c^2 C^2 D,$$

hence, dividing by 2, we have

$$2^{\alpha-1} e^2 E^2 - 2^{2k-\alpha-1} c^2 C^2 D = u.$$

Notice that we can exclude that $\alpha = 2k - 1$, because otherwise D would be a square modulo 2, contradicting the hypothesis. So $\alpha = 1$ and the equation reduces to

$$e^2 E^2 - 2^{2k-2} c^2 C^2 D = u. \tag{3.17}$$

Notice that, if u is a square in $\mathbb{Q}$, then $(eE/\sqrt{u}, 2^{k-1}cC/\sqrt{u})$ is again a solution of the Pell equation (3.16) for D, with $\deg E < \deg A$. But A was chosen to have minimal degree, which gives a contradiction.

Therefore, we can assume that u is not a square in $\mathbb{Q}$. We can also assume that $2k-2 \geq 2$, otherwise D would be a square modulo 2, contradicting the hypothesis.

We can then rewrite D as

$$D = \frac{e^2 E^2 - u}{2^{2k-2} c^2 C^2}. \tag{3.18}$$

As u is odd and $D \in \mathbb{Z}[t]$, we have that $4 \mid (e^2 E^2 + u)$. In particular, this means that all coefficients of monomials of positive degree in E are even and, as u is odd, the constant term of $e^2 E^2$ is congruent to 1 modulo 4. This means that $u \equiv 1 \mod 4$. If we factorize D, we obtain

$$D = \frac{(eE + \sqrt{u})(eE - \sqrt{u})}{2^{2k-2} c^2 C^2},$$

which gives a nontrivial factorization in $\mathbb{Q}(\sqrt{u})$ as u is not a square in $\mathbb{Q}$. Notice that this also gives a nontrivial factorization in $\mathbb{Q}_2(\sqrt{5})$ because if u is congruent to 1 modulo 8, then u is a square in $\mathbb{Q}_2$, while if u is congruent to 5 modulo 8, then it is a square in $\mathbb{Q}_2(\sqrt{5})$, contradicting the hypothesis that D is irreducible over $\mathbb{Q}_2(\sqrt{5})$. This proves the proposition. $\qquad\square$

BIBLIOGRAPHY

[1]   Y. André, P. Corvaja and U. Zannier: *The Betti map associated to a section of an abelian scheme (with an appendix by Z. Gao)*, arXiv: 1802.03204 (2018).

[2]   F. Barroero and L. Capuano: *Unlikely Intersections in families of abelian varieties and the polynomial Pell equation*, arXiv:1801.02885 (2018), 28 pp, submitted.

[3]   E. Bombieri and P. B. Cohen: *Siegel's lemma, Padé approximations and Jacobians*, Ann. Sc. Norm. Super. Pisa, Cl. Sci., IV. Ser. **25** (1997), no. 1–2, 155-178.

[4]   D. Bertrand: *Unlikely intersections in Poincaré biextensions over elliptic schemes*, Notre Dame J. Form. Log. **54** (2013), no. 3–4, 365-375.

[5]   D. Bertrand: *Generalized jacobians and Pellian polynomials*, J. Thor. Nombres Bordeaux **27** (2015), no. 2, 439-461.

[6]   D. Bertrand, D. Masser, A. Pillay and U. Zannier: *Relative Manin-Mumford for semi-abelian surfaces*, Proc. Edinb. Math. Soc. (2016), no. 4, 837-875.

[7]   P. Corvaja, D. Masser and U. Zannier: *Sharpening Manin-Mumford for certain algebraic groups of dimension 2*, Enseign. Math. (with a letter of Serre to Masser as an appendix), **59** (2013), no. 3–4, 225-269.

[8]   P. Corvaja, D. Masser and U. Zannier: *Torsion hypersurfaces on abelian schemes and Betti coordinates*, Math. Ann. 371 (2018), no. 3-4, 1013-1045.

[9]   J.H.E. Cohn: *The length of the period of the simple continued fraction of* $\sqrt{d}$, Pacific J. Math. **71** (1977), 21-32.

[10] L. Capuano, F. Veneziano and U. Zannier: *An effective criterion for periodicity of ℓ-adic continued fractions*, Math. Comp. 88 (2019), no. 318, 1851-1882.

[11] L. Euler: *De usu novi algorithmi in problemate pelliano solvendo*, Novi Commentarii Acad. Sci. Petropol. **11** (1767), 29-66.

[12] R. Hooke: *Linear p–adic groups and their Lie algebras*, Annals of Math. **43(4)** (1942), 641-655.

[13] A. Ya. Khinchin: *Continued fractions*, Dover Pub. (1997).

[14] S. Lang: *Division points on curves*, Ann. Mt. Pura Appl. **70** (1965), no. 4, 229-234.

[15] M. Laurent: *Équations diophantiennes exponentielles*, Invent. Math. **78** (1994), no. 2, 299-327.

[16] H. W. Lenstra, Jr.:*Solving the Pell equation*, Notices of the AMS **94** (2002), no. 2, 182-192.

[17] K. Mahler: *Zur approximation P-adischer irrationalzahlen*, Nieuw Arch. Wisk. **18** (1934), 22-34.

[18] F. Malagoli: *Continued fractions in function fields: polynomial analogues of McMullen's and Zaremba's conjectures*, PhD Thesis (2017), 131 pp, arXiv:1704.02640.

[19] D. Masser and U. Zannier: *Torsion Points, Pell's equation, and integration in elementary terms*, preprint (2018), 52 pp, submitted.

[20] J. McLaughlin: *Polynomial solutions of Pell's equation and fundamental units in real quadratic fields*, J. London Math. Soc. **67** (2003), no. 1, 16-28.

[21] C. McMullen: *Uniformly Diophantine numbers in a fixed real quadratic field.*, Compos. Math. **145** (2009), no. 4, 827-844.

[22] O. Merkert: *Reduction and specialiation of hyperelliptic continued fractions*, PhD Thesis (2017), 142 pp, arXiv:1706.04801.

[23] D. Masser and U. Zannier: *Torsion points on families of simple abelian surfaces and Pell's equation over polynomial rings*, J. Eur. Math. Soc. (JEMS) **17** (2015), no. 9, 2379-2416, with an appendix by E. V. Flynn.

[24] M. B. Nathanson: *Polynomial Pell's eeuations*, Proceedings of the AMS **56** (1976), 89-92.

[25] V. P. Platonov: *Number-theoretic properties of hyperelliptic fields and the torsion problem in Jacobians of hyperelliptic curves over the rational number field*, Uspekhi Mat. Nauk 69 **415** (2014), no. 1, 3–38.

[26] A. J. van der Poorten and X. C. Tran: *Quasi-elliptic integrals and periodic continued fractions*, Monatsch. Math. **131** (2000), 155–169.

[27] M. Rosenlicht: *Generalised Jacobian varieties*, Annals of Math. **59** (1954), 505-530.

[28] M. Ru: *A weak effective Roth's theorem over function fields*, Rocky Mountain J. of Math. **30** (2000), 723-734.

[29] A. Ruban: *Certain metric properties of the p-adic numbers*, Sibirsk. Mat. Z. **11** (1970), 222-227.

[30] P. Sarnak and S. Adams: *Betti numbers of congruence groups* (with an appendix by Zeev Rudnick, Israel J. Math. **88** (1994), nos. 1-3, 31-72.

[31] H. Schmidt: *Multiplication polynomials and relative Manin-Mumford*, PhD Thesis (2015).

[32] J. P. Serre: *Groupes algébriques et corps de classes*, Actualites Sci. et Ind., Hermann, Paris (1975).

[33] J. P. Serre: *Algebraic groups and class fields*, Springer-Verlag GTM **117** (1988).

[34] J. Silverman: *Heights and the specialization map for families of abelian varieties*, J. Reine Angew. Math. 342 (1983), 197-211.

[35] J. Sondow: *Irrationality measures, irrationality bases, and a theorem of Jarnik*, preprint, arXiv:0406300 (2004).

[36] *SageMath, the Sage mathematics software system* (Version 6.5), Sage Developers (2016), http://www.sagemath.org.

[37] S. Uchiyama: *Rational approximation to algebraic functions*, J. Fac. Sci. Hokkaido Univ. **15** (1961), 173-192.

[38] A. J. van der Poorten: *Formal power series and their continued fraction expansion*, Algorithmic number theory (Portland, OR, 1998), Lecture Notes in Comput. Sci., vol. 1423, Springer, Berlin (1998), 358-371.

[39] A. J. van der Poorten: *Reduction of continued fractions of formal power series*, Continued fractions: from analytic number theory to constructive approximation (Columbia, MO, 1998), Contemp. Math., vol. 236, Amer. Math. Soc., Providence, RI (1999), 343-355.

[40] A. J. van der Poorten: *Non-periodic continued fractions in hyperelliptic function fields*, Bull. Austral. Math. Soc. **64** (2001), no. 2, 331-343.

[41] J. Wang: *An effective Roth's theorem for function fields*, Rocky Mountain J. of Math. **26** (1996), 1225-1234.

[42] U. Zannier: *Lecture notes on Diophantine analysis*, Edizioni della Normale (2009). With an appendix by Francesco Amoroso.

[43] U. Zannier: *Some problems of unlikely intersections in arithmetic and geometry*, Annals of Math. Studies 181, Princeton University Press (2012), xiv+160 pp. With appendixes by David Masser.

[44] U. Zannier: *Unlikely intersections and Pell's equations in polynomials*, Trends in Contemporary Mathematics, Springer INdAM Series **8** (2014), 151-169.

[45] U. Zannier: *Hyperelliptic continued fractions and generalized Jacobians*, Amer. J. Math., 141 (2019), no.1, 1-40.

Chapter Four

Faltings Heights and L-functions: Minicourse Given by Shou-Wu Zhang

Ziyang Gao, Rafael von Känel, and Lucia Mocz

THESE ARE NOTES from the minicourse given by Shou-Wu Zhang (Princeton University). The notes were worked out and extended by Ziyang Gao, Rafael von Känel,[1] and Lucia Mocz.

This chapter essentially consists of three parts. In the first part (§4.1), we discuss conjectures and results in the literature which give bounds, or formulae in terms of L-functions, for "Faltings heights." We also mention various applications of such conjectures and results. The second part (§4.2 and §4.3) is devoted to the work of Yuan-Zhang [60] in which they proved the averaged Colmez conjecture. Here we discuss the main ideas and concepts used in their proof, and we explain in detail various constructions in [60]. In the third part (§4.4), we go into the function field world and we discuss the work of Yun-Zhang [61]. Therein they compute special values of higher derivatives of certain automorphic L-functions in terms of self-intersection numbers of Drinfeld-Heegner cycles on the moduli stack of shtukas. The result of Yun-Zhang might be viewed as a higher Gross-Zagier/Chowla-Selberg formula in the function field setting. In fact, throughout §4.3 and §4.4 we try to motivate and explain the philosophy that Chowla-Selberg type formulae (such as the averaged Colmez conjecture) are special cases of Gross-Zagier type formulae coming from identities between geometric and analytic kernels.

We tried to make this chapter accessible for non-specialists. In particular, we conducted quite some effort to add throughout explanations, discussions, background material, and references which were not part of the minicourse. We would like to thank Shou-Wu Zhang for answering questions, for giving useful explanations, and for sharing his insights.

---

[1] R. von Känel served as group leader of the working group "Minicourse Zhang" consisting of Javier Fresán, Ziyang Gao, Ariyan Javanpeykar, Rafael von Känel, Lucia Mocz, and Roland Paulin. The authors are grateful to Javier Fresán, Ariyan Javanpeykar, and Roland Paulin for their contribution to the draft version of this text.

4.1 HEIGHTS AND *L*-FUNCTIONS

The plan of this section is as follows. We begin in §4.1.1 with the definition of the Faltings height of abelian varieties over number fields. Then we discuss the generalized Szpiro conjecture and some of its applications, including the effective Mordell conjecture, no Siegel zeroes, and the abc-conjecture. In §4.1.2 we focus on the Faltings height of CM abelian varieties. After introducing some terminology from the theory of complex multiplication, we state and discuss the Colmez conjecture and its averaged version. Here we also mention two applications of these conjectures: The proof of the André-Oort conjecture for $\mathcal{A}_g$ and a new Northcott property for the Faltings height of CM abelian varieties. Finally, in §4.1.3, we briefly discuss the work of Yun-Zhang on higher derivatives of certain automorphic *L*-functions over function fields.

4.1.1 The Faltings Height and the Generalized Szpiro Conjecture

After defining the Faltings height, we discuss in this section selected aspects of the generalized Szpiro conjecture for abelian varieties. Let $g \geq 1$ be a rational integer, let K be a number field, and let A be an abelian variety defined over K of dimension g. We denote by $\mathcal{O}_K$ the ring of integers of K and we let $\mathcal{A}$ be the Néron model of A over $\mathcal{O}_K$ with zero section e.

The Faltings height. The Hodge bundle $\omega(\mathcal{A}) = c^* \Omega^g_{\mathcal{A}/\mathcal{O}_K}$ of $\mathcal{A}$ is a line bundle on $\mathrm{Spec}(\mathcal{O}_K)$. For Arakelov theoretic considerations, we endow $\omega(\mathcal{A})$ with the structure of a hermitian line bundle by equipping it with hermitian metrics $\overline{\omega}(\mathcal{A}) = (\omega(\mathcal{A}), \{\|\cdot\|_\sigma\}_{\sigma:K \hookrightarrow \mathbb{C}})$, where

$$\|\alpha\|_\sigma^2 = \frac{1}{(2\pi)^g} \left| \int_{A_\sigma(\mathbb{C})} \alpha \wedge \overline{\alpha} \right|, \quad \alpha \in \omega(\mathcal{A}) \otimes_\sigma \mathbb{C}, \qquad (4.1)$$

and where the subscript σ denotes the base change from K to $\mathbb{C}$ via an embedding $\sigma: K \hookrightarrow \mathbb{C}$. We point out that Faltings [19, §3] normalizes the canonical metric (4.1) via the factor 2^{-g}, while we use the factor $(2\pi)^{-g}$ following Deligne [14, p. 26] and Yuan-Zhang [60, §1].

Definition 4.1 (Faltings Height). We first give the definition in the special case when $\mathcal{A}$ is semi-abelian over $\mathcal{O}_K$. In this case the *Faltings height* $h(A)$ of A is defined by

$$h(A) = \frac{1}{[K:\mathbb{Q}]} \widehat{\deg} \, \overline{\omega}(\mathcal{A}) = \frac{1}{[K:\mathbb{Q}]} \left(\log |\omega(\mathcal{A})/\alpha \mathcal{O}_K| - \sum_{\sigma:K \hookrightarrow \mathbb{C}} \log \|\alpha\|_\sigma \right)$$

where $\widehat{\deg}$ denotes the arithmetic degree and where $\alpha \in \omega(\mathcal{A})$ is a global section, independent of choice by the adelic product formula. In general, we may and do take a finite field extension L of K such that the Néron model of $A_L = A \times_K L$ over $\mathcal{O}_L$ is semi-abelian and then we define $h(A) = h(A_L)$. This definition is independent of the choice of L as above, since the identity component of a semi-abelian Néron model commutes with base change.

While $h(A)$ does not measure the additive reduction of A, it has the advantage of being stable under base change in the following sense: For any finite field extension L of K, it holds that $h(A) = h(A \times_K L)$. In particular, $h(A)$ is invariant under geometric isomorphisms of A. In the literature $h(A)$ is often called the *stable* Faltings height of A. Faltings [19, §3] originally introduced this height in his proof of the Mordell, Shafarevich, and Tate conjectures. He showed inter alia the following finiteness theorem which is now known as the *Northcott property* for $h(A)$.

Theorem 4.2 (Northcott Property). *Let K be a number field, let $g \geq 1$ be an integer, and let $c > 0$ be a real number. Then there are only finitely many isomorphism classes of abelian varieties A over K of dimension g with semi-abelian reduction over $\mathcal{O}_K$ such that $h(A) \leq c$.*

In his original proof, Faltings first relates $h(A)$ to the Weil height of a point on the moduli space of (principally) polarized abelian varieties and then he deduces finiteness from the classical Northcott property of the projective space; here Zarhin [62] removed polarization assumptions. Faltings-Chai [22] later gave a conceptually simpler proof of Theorem 4.2.

Generalized Szpiro conjecture. Faltings's proof of the Mordell, Shafarevich, and Tate conjectures was inspired by works of Arakelov [4], Parshin [43], and Szpiro [48] on analogous problems for curves and abelian varieties over function fields. In the case of function fields of complex curves, one can use Hodge theory to prove the following so-called "Arakelov inequality."

Theorem 4.3 (Arakelov Inequality). *Let B be a smooth, projective, and connected curve over $\mathbb{C}$. Suppose that A is a semi-abelian scheme over B of relative dimension $g \geq 1$ such that A is abelian over a nonempty open subscheme $S_A \subseteq B$. If $-\chi(S_A) \geq 0$ and $T_A = B - S_A$, then*

$$\deg \omega(A) \leq \tfrac{g}{2}(|T_A| + 2g(B) - 2) = \tfrac{g}{2}(-\chi(S_A)). \tag{4.2}$$

Here $\chi(S_A)$ is the topological Euler-Poincaré characteristic of S_A in the curve B of genus $g(B)$, and $\omega(A) = e^*\Omega^g_{A/B}$ for e the zero section of A. The inequality (4.2) is sharp.

This theorem, which is essentially due to Faltings [20], generalizes earlier results of Arakelov [4] and Parshin [43]. See also Deligne [15] who refined parts of Faltings's arguments via Zucker's work [64]. Inspired by the Arakelov inequality for elliptic curves over function fields, Szpiro proposed in 1982 his "discriminant conjecture" which gives a version of (4.2) for elliptic curves over number fields. Szpiro's conjecture led to the essentially equivalent "abc-conjecture" of Masser-Oesterlé [38], which became a central open problem in number theory. The following conjecture implies the discriminant conjecture [49, Conj. 1] for elliptic curves over K, and it strengthens the discriminant conjecture [29, (D)] for arbitrary curves over K of genus ≥ 1.

Conjecture 4.4 (Generalized Szpiro Conjecture).   *Let K be a number field and let $g \geq 1$ be an integer. Then any abelian variety A over K of dimension g satisfies*

$$h(A) \leq \tfrac{\alpha}{[K:\mathbb{Q}]} \left(\frac{1}{g} \log N_A + 2 \log \Delta_K \right) + \beta, \tag{4.3}$$

where $\alpha, \beta \in \mathbb{R}$ are constants depending only on $[K:\mathbb{Q}]$ and g. Here Δ_K denotes the absolute discriminant of $K/\mathbb{Q}$ and N_A denotes the norm of the conductor ideal of A/K.

This conjecture might be viewed as a version of Frey's height conjecture [23, p. 39]. Further, we observe that (4.3) is an arithmetic analogue of the Arakelov inequality (4.2), with the arithmetic curve $\mathrm{Spec}(\mathcal{O}_K)$ used in place of the complex curve B and with $[K:\mathbb{Q}]h(A) = \widehat{\deg}\,\overline{\omega}(\mathcal{A})$, $\frac{1}{g} \log N_A$, and $\log D_K$ playing the role of $\deg \omega(A)$, $|T_A|$, and $g(B)$, respectively. The generalized Szpiro conjecture has many striking consequences, such as, for example:

  (i) **The effective Mordell conjecture:** If Conjecture 4.4 holds for K with α, β both effectively computable for each $g \geq 1$, then one obtains[2] an effective version of the Mordell conjecture: For any smooth, projective, and geometrically connected curve X of genus at least two defined over K, one can in principle determine all K-rational points of X.
 (ii) **The abc-conjecture:** If Conjecture 4.4 holds for K and $g = 1$, then the abc-conjecture over K holds with some fixed radical exponent.

The Arakelov inequality (4.2) suggests that one can take $\alpha = \frac{g}{2}$ in (4.3). However, on using, for example, the family of elliptic curves constructed by Masser [37] and the additivity of the Faltings height on products, we see that for each $g \geq 1$ Conjecture 4.4 cannot hold with $\alpha = \frac{g}{2}$. Therefore the following conjecture is optimal in terms of α.

Conjecture 4.5 (Strong Generalized Szpiro Conjecture).   *For each real number $\epsilon > 0$, the inequality (4.3) in the generalized Szpiro conjecture holds with*

$$\alpha = \frac{g}{2} + \epsilon \quad \text{and} \quad \beta = \beta(g, \epsilon) \tag{4.4}$$

where $\beta(g, \epsilon) \in \mathbb{R}$ is a constant which depends only on g and ϵ.

---

[2]This only requires an effective upper bound for $h(A)$ in terms of N_A, g for each $g \geq 1$ and $K = \mathbb{Q}$.

While the above discussion provides some motivation for taking $\alpha = \frac{g}{2} + \epsilon$, we are not aware of anything indicating that β can be chosen as conjectured in (4.4) and in fact we are rather skeptical about β being independent of $[K:\mathbb{Q}]$. However, the independence of α, β on $[K:\mathbb{Q}]$ assures that a version of Conjecture 4.5 for CM elliptic curves has some striking applications in analytic number theory which we shall discuss below. For these applications it is essentially irrelevant how the conjectured bound depends on N_A. This is somewhat surprising, since usually the dependence on N_A is crucial for applications of the generalized Szpiro conjecture. Finally, we remark that one also might call the above Conjecture 4.5 the "Uniform Generalized Szpiro Conjecture," since α and β are both independent of the degree $[K:\mathbb{Q}]$.

4.1.2   CM Abelian Varieties and Their Faltings Height

In this section, we first recall some basic terminology and definitions from the theory of complex multiplication. Then we consider the CM case of the generalized Szpiro conjecture and we discuss an application to Siegel zeroes. Finally we state the Colmez conjecture and its averaged version, and we mention some recent applications of these conjectures.

Theory of complex multiplication.   We now give a brief survey of the relevant theory of complex multiplication, and refer to [8] for a comprehensive treatment. A *CM field* E is a totally imaginary quadratic extension of a totally real number field F. The nontrivial element $c \in \mathrm{Gal}(E/F)$ is called the *CM involution* on E. A *CM type* Φ of a CM field E is a subset $\Phi \subseteq \mathrm{Hom}(E, \mathbb{C})$ such that $\Phi \sqcup \Phi \circ c = \mathrm{Hom}(E, \mathbb{C})$, where Hom means field morphisms. The fixed field of c is the totally real subfield $F \subseteq E$ which satisfies $[E:F] = 2$. We call F the *maximal totally real subfield* of E. A *CM algebra* P is a finite nonempty product $P = \prod E_i$ of CM fields E_i and a *CM type* Φ of P is the disjoint union $\Phi = \sqcup_i \Phi_i$ of CM types Φ_i of E_i.

There are several non-equivalent notions of a CM abelian variety in the literature. To explain what we mean by a CM abelian variety (defined over a subfield of $\mathbb{C}$), we denote by $\mathrm{End}(B)$ the endomorphism ring of the S-group scheme underlying an arbitrary abelian scheme B over an arbitrary base scheme S and we write $\mathrm{End}^0(B) = \mathrm{End}(B) \otimes_{\mathbb{Z}} \mathbb{Q}$. Let A be an abelian variety of dimension $g \geq 1$ defined over a subfield of $\mathbb{C}$, and denote by $A_{\mathbb{C}}$ the base change of A to $\mathbb{C}$. Further let P be a CM algebra with $[P:\mathbb{Q}] = 2g$ and let Φ be a CM type of P.

Definition 4.6.   We say that A has *CM by* (P, Φ) if the following two conditions hold:

(i) There is an injective $\mathbb{Q}$-algebra homomorphism $\iota : P \hookrightarrow \mathrm{End}^0(A_{\mathbb{C}})$.
(ii) The $P \otimes_{\mathbb{Q}} \mathbb{C}$-module structure of $\mathrm{Lie}(A_{\mathbb{C}})$ induced by ι determines Φ.

In the case when A has CM by (P, Φ), we also say that A *has CM by* $(E, \sum_i \Phi_i)$ when in the product $P = \prod E_i$ all E_i are the same CM field E and we also say that A has *CM by* $(\mathfrak{a}, \Phi)$ when $\mathfrak{a}$ is a subset of P such that $\iota(\mathfrak{a}) \subseteq \mathrm{End}(A_\mathbb{C})$. If (i) holds for some ι, then we say that A has *CM by* P and we call the pair (A, ι) a *CM abelian variety*; we shall write $A = (A, \iota)$ in situations where the specific choice of ι is irrelevant. Finally, if (A, ι) is a CM abelian variety and $\mathfrak{a}$ is a subset of P such that $\iota(\mathfrak{a}) \subseteq \mathrm{End}(A_\mathbb{C})$, then we say that A has *CM by* $\mathfrak{a}$.

Remark 4.7. Conditions (i) and (ii) of the above definition are of geometric nature, since they only depend on $A_\mathbb{C}$. Further, our notion of a CM abelian variety is slightly more general than [8, 1.3.8.1] which uses A in place of $A_\mathbb{C}$. For example, if A is an elliptic curve over $\mathbb{Q}$ with $\mathrm{End}(A_\mathbb{C}) \neq \mathbb{Z}$, then A is a CM abelian variety in our sense but never in the sense of [8].

We next review some basic properties of CM abelian varieties. To simplify the discussion, we assume from now on that $P = E$ is a CM field. If A has CM by E, then the complex abelian variety $A_\mathbb{C}$ is isotypic which means that $A_\mathbb{C}$ is isogenous to a power of a simple complex abelian variety. Further, there exists a complex abelian variety A_Φ with CM by $(\mathcal{O}_E, \Phi)$ such that

$$A_\Phi(\mathbb{C}) = \mathbb{C}^g / \Phi(\mathcal{O}_E). \tag{4.5}$$

Here $\Phi(\mathcal{O}_E)$ denotes the lattice of $\mathbb{C}^g$ given by the image of $\mathcal{O}_E$ under the injective map $E \hookrightarrow \mathbb{C}^g$ defined by the g-distinct embeddings $E \hookrightarrow \mathbb{C}$ in Φ. The complex abelian variety A_Φ is isotypic but not necessarily simple. We recall that our A is defined over an arbitrary subfield of $\mathbb{C}$. In the case when A is a CM abelian variety, one knows that A descends to an abelian variety A_K defined over a number field K and then we define the Faltings height of A by $h(A) = h(A_K)$. This does not depend on the choice of K, since the stable Faltings height is invariant under geometric isomorphisms. Further we define the Faltings height $h(\Phi)$ of Φ by

$$h(\Phi) = h(A_\Phi). \tag{4.6}$$

This notation is motivated by a result of Colmez [13, Thm. 0.3 (ii)] which implies that any complex abelian variety A' with CM by $(\mathcal{O}_E, \Phi)$ satisfies $h(\Phi) = h(A')$. Finally, any CM abelian variety defined over a number field has potentially good reduction everywhere.

The generalized Szpiro conjecture for CM abelian varieties. For any CM abelian variety one can essentially ignore the contribution of its conductor in the generalized Szpiro conjecture, since it has potentially good reduction everywhere. This motivates the following:

Conjecture 4.8. *Let K be a number field and let $g \geq 1$ be an integer. Then any CM abelian variety A over K of dimension g, which extends to an abelian scheme over $\mathcal{O}_K$, satisfies*

$$h(A) \leq \tfrac{\gamma}{[K:\mathbb{Q}]} \log \Delta_K + \delta, \tag{4.7}$$

where $\gamma, \delta \in \mathbb{R}$ are constants which depend only on g.

We notice that Conjecture 4.5 directly implies a stronger version of inequality (4.7) for any (not necessarily CM) abelian variety over a number field K which extends to an abelian scheme over $\mathcal{O}_K$. Although the conductor aspect disappears, the above conjecture is still of interest. To illustrate this, we now consider the case of CM elliptic curves and we let E be an imaginary quadratic field of discriminant $-d < 0$. Class field theory gives that any complex elliptic curve with CM by $\mathcal{O}_E$ is defined over the Hilbert class field H of E, and it holds that

$$\tfrac{1}{h} \log \Delta_H = \log \Delta_E = \log d$$

where $h = [H : E]$ is the class number of $\mathcal{O}_E$. Then Conjecture 4.5 motivates the following conjecture which one might call "strong generalized Szpiro conjecture for CM elliptic curves."

Conjecture 4.9. *Suppose that $\epsilon > 0$ is a real number. Then, for any imaginary quadratic field E of discriminant $-d < 0$ and for each complex elliptic curve A with CM by $\mathcal{O}_E$, it holds*

$$h(A) \le (1/2 + \epsilon) \log d + c_\epsilon$$

where $c_\epsilon \in \mathbb{R}$ is a constant which depends only on ϵ.

We now discuss an application of this conjecture. Let $\eta : (\mathbb{Z}/d\mathbb{Z})^\times \to \{\pm 1\}$ be the quadratic character of our imaginary quadratic number field E of discriminant $-d < 0$ and denote by $L(\eta, s)$ the Dirichlet L-function of η. Then we consider the following conjecture.

Conjecture 4.10 (No Siegel Zeroes). *The Dirichlet L-function $L(\eta, s)$ has no real zero in the interval $1 - \frac{c}{\log d} \le s \le 1$ for some sufficiently small absolute constant $c > 0$.*

This conjecture is of fundamental interest in number theory. To explain a connection to the generalized Szpiro conjecture for CM elliptic curves, we consider the following reformulation of the classical Lerch-Chowla-Selberg formula in terms of the Faltings height; see §4.3.2 for a discussion of this formula which can be deduced from Kronecker's limit formula.

Theorem 4.11 (Lerch-Chowla-Selberg Formula). *We denote by $L'(\eta, s)$ the derivative of the Dirichlet L-function $L(\eta, s)$. Then any complex elliptic curve A with CM by $\mathcal{O}_E$ satisfies*

$$h(A) = -\tfrac{1}{2} \frac{L'(\eta, 0)}{L(\eta, 0)} - \tfrac{1}{4} \log d. \tag{4.8}$$

Now, on combining this formula with analytic number theory arguments (which can be found, for example, in Granville-Stark [26]; see also Chowla [9,10] and Mahler [35]), one sees that Conjecture 4.9 implies Conjecture 4.10 on the non-existence of Siegel zeroes of $L(\eta, s)$.

Remark 4.12 (Euler's "numeri idonei"). The arguments in [26] are effective and can be used to show, moreover, the following: If the constant c_ϵ in Conjecture 4.9 is effectively computable for each $\epsilon > 0$, then one can, in principle, determine all of Euler's "numeri idonei."

To conclude our discussion of the generalized Szpiro conjecture, we point out that there are many important aspects of this conjecture which we could not discuss in this short section. For instance, we did not mention several interesting reformulations, function field analogues, and applications of the case $g = 1$ of Conjectures 4.4 and 4.5 which can be found in the extensive literature on the abc-conjecture. Finally, we would like to thank Vesselin Dimitrov for useful discussions on the various Szpiro conjectures and their applications, for answering several questions, and for sending us many interesting and insightful comments.

The Colmez conjecture. In the literature there are several generalizations and reformulations (such as Theorem 4.11) of the classical Lerch-Chowla-Selberg formula. For example, Gross [27] gave a new geometric proof and together with Deligne he conjectured a formula (modulo algebraic numbers) for the periods of certain CM motives. Anderson [1] reformulated the Gross-Deligne conjecture in terms of the logarithmic derivative of an L-function at $s = 0$. Then Colmez [13] conjectured a precise formula for the Faltings height of CM abelian varieties and not only for their archimedean periods. He also motivated his conjecture by studying numerical relations satisfied by the product of the local periods of CM abelian varieties.

To state the Colmez conjecture, we follow the exposition in [56, §3]. Consider the Galois group $G = \mathrm{Gal}(\mathbb{Q}^{\mathrm{CM}}/\mathbb{Q})$ of the compositum $\mathbb{Q}^{\mathrm{CM}}$ of all CM fields inside $\overline{\mathbb{Q}}$, where $\overline{\mathbb{Q}}$ denotes the algebraic closure of $\mathbb{Q}$ inside $\mathbb{C}$. Let $E \subseteq \mathbb{C}$ be a CM field. To any CM type Φ of E, Colmez attached a locally constant class function A_Φ^0 on G with values in $\overline{\mathbb{Q}}$; we refer to [56, p. 465] for a concise construction of this function. Brauer's theorem gives that

$$A_\Phi^0 = \sum_\chi a_\chi \chi \tag{4.9}$$

can be written as a linear combination of irreducible Artin characters χ of G, where the numbers a_χ are zero for almost all χ. For any Artin character χ of G, we denote by $L'(\chi, s)$ the derivative of the Artin L-function $L(\chi, s)$ associated to χ and we write f_χ for the analytic Artin conductor of χ. Now we can state the Colmez conjecture [13, Conj. 0.4, Thm. 0.3].

Conjecture 4.13 (Colmez Conjecture). *Let Φ be a CM type of an arbitrary CM field $E \subseteq \mathbb{C}$. If $A_\Phi^0 = \sum_\chi a_\chi \chi$ is the decomposition (4.9) of the function A_Φ^0 attached to Φ, then it holds*

$$h(\Phi) = -\sum_\chi a_\chi \left(\tfrac{L'(\chi, 0)}{L(\chi, 0)} + \tfrac{1}{2} \log f_\chi \right) + \tfrac{1}{4} \log 2\pi.$$

We point out that our normalization of the Faltings height differs from the one used by Colmez. In the case when E is quadratic, the Colmez conjecture is equivalent to the reformulation in Theorem 4.11 of the classical Lerch-Chowla-Selberg formula. On using and refining the work of Gross [27] and Anderson [1], Colmez [13] proved his conjecture in the case when $E/\mathbb{Q}$ is abelian; in fact, Colmez originally proved the result only up to rational multiples of $\log 2$ which were then later removed by Obus [42]. Further, Yang [57,58] introduced a new method which he used to establish the Colmez conjecture in the case when $[E:\mathbb{Q}] \leq 4$.

Averaged Colmez conjecture. Let E be a CM field with maximal totally real subfield F and write $g = [F:\mathbb{Q}]$. Colmez [13, p. 634] also formulated an averaged version of his conjecture giving an expression independent of a specific CM type. In the case $g = 1$ the averaged version is equivalent to the Colmez conjecture. This allows us to see the following result as another generalization of the classical Lerch-Chowla-Selberg formula.

Theorem 4.14 (Averaged Colmez Conjecture). *Let $\Delta_{E/F}$ be the norm of the relative discriminant of E/F, and let Δ_F be the absolute discriminant of F. Then it holds*

$$\frac{1}{2^g} \sum_{\Phi} h(\Phi) = -\frac{1}{2} \frac{L'(\eta, 0)}{L(\eta, 0)} - \frac{1}{4} \log(\Delta_{E/F} \Delta_F)$$

with the sum taken over all distinct CM types Φ of E. Here $L'(\eta, s)$ is the derivative of the L-function $L(\eta, s)$ of the quadratic character η of $\mathbb{A}_F^{\times}/F^{\times}$ defined by E/F.

The above theorem was proven by Yuan-Zhang [60] and independently by Andreatta-Goren-Howard-Madapusi Pera [3]. While both proofs studied the geometry of Shimura varieties, they are very different. The proof in [3] involves higher dimensional Shimura varieties and it applies the method of Yang [57,58] which he introduced to prove the Colmez conjecture for $g \leq 2$. On the other hand, the proof in [60] involves Shimura curves and it uses the method of Yuan-Zhang-Zhang [59] which they developed to prove generalized Gross-Zagier formulas for Shimura curves. More precisely, the proof of Yuan-Zhang consists of two parts. In the first part, they consider a certain quaternionic Shimura curve X and they show that

$$\frac{1}{2^g} \sum_{\Phi} h(\Phi) = \frac{1}{2} h(P) - \frac{1}{4} \log(\Delta_F \Delta_{\mathbb{B}}), \qquad (4.10)$$

where $h(P)$ is the height of a CM point $P \in X$ and $\Delta_{\mathbb{B}}$ is the discriminant of the quaternion algebra $\mathbb{B}$ associated to X. We refer to §4.2 for a precise statement of (4.10) and for a detailed discussion of the strategy of proof of (4.10) which combines various new ideas and concepts. In the second part, they establish the following generalized Chowla-Selberg formula:

$$\tfrac{1}{2}h(P) = -\tfrac{1}{2}\frac{L'(\eta,0)}{L(\eta,0)} - \tfrac{1}{4}\log(\Delta_{E/F}/\Delta_{\mathbb{B}}). \qquad (4.11)$$

In §4.3 we shall discuss the proof of (4.11) which uses an extension of the method of Yuan-Zhang-Zhang [59]. In fact Yuan proved (4.11) already in 2008, while Zhang showed (4.10) in 2015. Finally, the averaged Colmez conjecture in Theorem 4.14 is a direct consequence of the averaged height formula in (4.10) and the generalized Chowla-Selberg formula in (4.11).

Applications of the Colmez conjectures.   The averaged Colmez conjecture played a key role in the recent proof of the André-Oort conjecture for $\mathcal{A}_g$ by Tsimerman [50]. The latter conjecture asserts that any irreducible component of the Zariski closure of any set of CM points in $\mathcal{A}_g$ is a special subvariety of $\mathcal{A}_g$, meaning that it comes from a sub-Shimura datum (up to Hecke translation). We refer to [50] for details about this conjecture.

Next, we discuss some aspects of the proof of the André-Oort conjecture for $\mathcal{A}_g$ in order to explain the role of the averaged Colmez conjecture. Based on his joint work with Zannier on the Manin-Mumford conjecture [47], Pila [44] used o-minimal theory (especially the Pila-Wilkie counting theorem [46]) to give the first unconditional proof of the André-Oort conjecture for $Y(1)^N$.[3] This method, known as the "Pila-Zannier strategy," was then generalized by Ullmo-Yafaev [52] to projective Shimura varieties, Pila-Tsimerman [45] to $\mathcal{A}_g$, and Klingler-Ullmo-Yafaev [30] to an arbitrary Shimura variety. Hence, the André-Oort conjecture was reduced to a lower bound on the Galois orbits of special points; see [51]. For $\mathcal{A}_g$, this lower bound is the following conjecture proposed by Edixhoven [18] and recently proven by Tsimerman [50].

Theorem 4.15 (Tsimerman).   *For any integer $g \geq 1$, there exist two constants $c_g, \delta_g > 0$ such that the following condition holds. For any CM field E of degree $2g$ and primitive CM type Φ for E, if A is an abelian variety of dimension g having CM by $(\mathcal{O}_E, \Phi)$, then*

$$[K:\mathbb{Q}] \geq c_g |\Delta_E|^{\delta_g}$$

for any number field K such that A is defined over K.

A main ingredient for this proof was the averaged Colmez conjecture which was used to obtain the following Brauer-Siegel bound: $h(\Phi) \ll_\epsilon |\Delta_E|^\epsilon$ for any $\epsilon > 0$. Combined with the Masser-Wüstholz isogeny theorem [39], this bound shows Theorem 4.15. We remark that the Pila-Zannier strategy was extended by Gao [24, 25] to mixed Shimura varieties, so Theorem 4.15 also implies the André-Oort conjecture for all mixed Shimura varieties of abelian type.

---

[3]The only unconditional result before Pila is André's proof for $N = 2$ [2].

Remark 4.16 (Effectivity). Although the averaged Colmez conjecture in Theorem 4.14 is explicit in all aspects and thus effective, the above described proof of the André-Oort conjecture is currently not known to be effective in general. However, in the literature there are effective proofs of special cases of the André-Oort conjecture. For example, Kühne [32, 33] and independently Bilu-Masser-Zannier [6] obtained effective proofs of the André-Oort conjecture for $Y(1)^2$; see also Wüstholz [55], Bilu-Luca-Masser [5], and the references therein.

Another application of the Colmez conjecture was the work by Mocz [40] to show a new Northcott property for the Faltings height. Her theorem is the following.

Theorem 4.17 (CM Northcott Property, [40]). *Fix $C \in \mathbb{R}_{>0}$ and $g \in \mathbb{Z}_{>0}$.*

1. *The number of isomorphism classes of CM abelian varieties A of dimension g with $h(A) < C$ is finite within $\overline{\mathbb{Q}}$-isogeny classes.*
2. *Assuming the Colmez conjecture and the Artin conjecture, the number of $\overline{\mathbb{Q}}$-isomorphism classes of CM abelian varieties A of dimension g with $h(A) < C$ is finite.*

In the first part, the author develops new techniques using (integral) p-adic Hodge theory to compute the differences in Faltings height between elements in the same isogeny class. These computations also allow the author to write new Colmez-type formulae for the Faltings height of all CM abelian varieties, not just those which are simple with CM by a maximal order. Some instances are worked out explicitly in [40].

It should be remarked that the CM Northcott property differs in character from the Northcott property shown by Faltings in Theorem 4.2 in the following two ways. First, no moduli space techniques or comparison of heights theorem is necessary in the proof, so it may be viewed as a more intrinsic Northcott property to the Faltings height. Second, unlike Faltings's Northcott property, the CM Northcott property counts abelian varieties over number fields of varying degree which can grow to be arbitrarily large.

4.1.3 The Function Field Setting: The Work of Yun-Zhang

In a parallel setting to the previous developments, Yun-Zhang [61] made a significant breakthrough to express the Taylor expansion of a certain class of L-functions over a function field $F = k(X)$ for a curve X over a finite field k of characteristic $p > 2$. The L-function in consideration was attached to a cuspidal automorphic representation of $G = \mathrm{PGL}_{2,F}$. The result is a simultaneous generalization of both the Waldspurger and Gross-Zagier formulae in the number field setting, which correspond to the special values computed in the first two terms of the Taylor expansion, respectively.

A brief introduction to shtuka. The moduli space of shtuka, Sht_G^r, provides the geometric setting for the intersection problem considered in the function field setting. The existence of such a moduli space, however, is strikingly different from its analogous spaces in the number field setting. In particular, the moduli space Sht_G^r can be defined for any integer $r \geq 0$ and admits a natural fibration over the r-fold self-product X^r of the curve X over k:

$$\mathrm{Sht}_G^r \to X^r.$$

In the number field case, the analogous spaces only exist (at least at the time this article was written) when $r \leq 1$.

A (balanced) GL_n-shtuka is the following object.

Definition 4.18. Let S be a scheme over X^r. A (balanced) GL_n-shtuka on S is a diagram of vector bundles of rank n on $X \times S$:

$$\underline{\mathcal{E}} = \left(\mathcal{E}_0 \xrightarrow{f_1} \mathcal{E}_1 \xrightarrow{f_2} \ldots \xrightarrow{f_{\frac{r}{2}}} \mathcal{E}_{r/2} \xleftarrow{f_{\frac{r}{2}+1}^{-1}} \mathcal{E}_{r/2+1} \xleftarrow{f_{\frac{r}{2}+2}^{-1}} \ldots \xleftarrow{f_r^{-1}} \mathcal{E}_r \simeq \mathcal{E}_0^\sigma \right)$$

where ${}^\sigma\mathcal{E}_0 := (1 \times \mathrm{Frob}_S)^*\mathcal{E}_0$ and such that the subquotient bundles

$$\mathcal{E}_i/\mathcal{E}_{i-1} \text{ for } 1 \leq i \leq r/2$$
$$\mathcal{E}_i/\mathcal{E}_{i+1} \text{ for } r/2 \leq i \leq r-1$$

take the form $(\Gamma_{x_i})_*\mathcal{L}_i$, where for each i we have that $x_i : S \hookrightarrow X$ is a morphism of schemes, $\Gamma_{x_i} \subseteq X \times S$ is the graph of x_i, and $\mathcal{L}_i$ is a line bundle on S.

The moduli stack of such objects is denoted by Sht_n^r. The objects parametrized by Sht_G^r will be GL_2-shtuka quotiented by an equivariant $\mathrm{Pic}(k)$-action on each object in the diagram, i.e., $\mathrm{Sht}_G^r := \mathrm{Sht}_2^r/\mathrm{Pic}_X(k)$. For more general definitions of shtuka see §4.4.2.

Heegner-Drinfeld cycles and the intersection problem. Let $\nu : X' \to X$ be a non-trivial étale double cover of X and $F' = k(X')$ its corresponding function field. We have the following commutative diagram

$$\begin{array}{ccc} \mathrm{Sht}_{1,X'}^r & \longrightarrow & \mathrm{Sht}_{2,X}^r \\ \downarrow & & \downarrow \\ X'^r & \xrightarrow{\nu^r} & X^r. \end{array}$$

Here, $\mathrm{Sht}_{1,X'}^r := \mathrm{Sht}_1^r \times_X^r X'^r$ and the top horizontal arrow is induced by the pushforward

$$\mathcal{L} \mapsto \nu_*\mathcal{L}.$$

Let $T = \mathrm{Res}_{F'/F}\mathbb{G}_m/\mathbb{G}_m \hookrightarrow G$ and define $\mathrm{Sht}_T^r := \mathrm{Sht}_{1,X'}^r/\mathrm{Pic}_X(k)$. This is a torsor over X' under $\mathrm{Pic}_{X'}(k)/\mathrm{Pic}_X(k)$. For more details, see §4.4.2.

There is a natural map

$$\mathrm{Sht}_T^r \to \mathrm{Sht}_G'^r := \mathrm{Sht}_G^r \times_{X^r} X'^r.$$

This defines a cycle of codimension r and is hence a representative of a class $\mathcal{Z} \in \mathrm{Ch}_c^r(\mathrm{Sht}_G'^r)_{\mathbb{Q}}$.

Theorem 4.19 (Yun-Zhang [61]). *The intersection number*

$$h(Z) =: \langle Z, Z \rangle = c_r(\Omega_{M/X^r})[Z]$$

is well-defined and equals

$$\frac{2^{r+2}}{(\log q)^r} L^{(r)}(\eta, 0).$$

Remark 4.20. When $r = 0$ this returns the class number formula, and when $r = 1$ we obtain a function field version of the Chowla-Selberg formula.

This theorem is stated precisely in Theorem 4.32, Theorem 4.47, and Theorem 4.48.

Significance and open questions. Prior to this result, it was a commonly held belief that only the central vanishing order of motivic L-functions carry geometric meaning; this result sheds light that there may be a geometric way to calculate the r-th Taylor coefficient for any $r \geq 0$ even if at present a precise conjecture cannot be formulated.

The relation to heights expressed by the first two Taylor coefficients also prompt us to ask whether there are other function field-theoretic analogues of the previous classical theorems that we considered. In particular: What is the right formulation of Szpiro's conjecture? What is the right formulation of the Mordell conjecture? What arithmetic questions hold in the function field setting for the function field analogues of CM points?

4.2 SHIMURA CURVES AND AVERAGED COLMEZ CONJECTURE

In this section we discuss the result in Yuan-Zhang [60] which relates the averaged height of CM abelian varieties with the height of CM points on a certain quaternionic Shimura curve. This result reduces in particular the averaged Colmez conjecture to the generalized Chowla-Selberg formula considered in the next section. After stating the result, we explain the main ingredients and concepts used in the proof which is outlined in §4.2.1 below.

To state the result, we take an integer $g \geq 1$ and we let E be a CM field of degree $[E : \mathbb{Q}] = 2g$ with maximal totally real subfield $F \subseteq E$. Let $\mathbb{B}$ be a totally definite incoherent quaternion algebra over $\mathbb{A}_F$. Then the group $\mathbb{B}_f^{\times}$ acts from the right on the projective limit $X = \lim_U X_U$ of the quaternionic Shimura curves X_U associated in §4.2.6 to open compact subgroups U of $\mathbb{B}_f^\times$. Here $X_U = X/U$ is inter alia a smooth and projective curve over F. For any point $P \in X$ we denote by P_U its image in X_U, and for any CM type Φ of E we denote by $h(\Phi)$ its Faltings height defined in (4.6). Now we can state [60, Thm. 1.6].

Theorem 4.21. *Suppose that there exists an embedding $\mathbb{A}_E \hookrightarrow \mathbb{B}$ over $\mathbb{A}_F$, and assume that $U = \prod U_v$ is a maximal open compact subgroup of $\mathbb{B}_f^\times$ containing $\widehat{\mathcal{O}}_E^\times$. If there is no finite place of F which ramifies both in E and $\mathbb{B}$, and $P \in X$ is fixed by $E^\times$, then*

$$\frac{1}{2^g} \sum_\Phi h(\Phi) = \tfrac{1}{2} h_{\overline{\mathcal{L}}_U}(P_U) - \tfrac{1}{4} \log(\Delta_\mathbb{B} \Delta_F). \tag{4.12}$$

Here the sum is taken over all distinct CM types Φ of E, Δ_F denotes the absolute discriminant of $F/\mathbb{Q}$, and $\Delta_\mathbb{B}$ denotes the norm of the product of the prime ideals of F over which $\mathbb{B}$ ramifies.

In the above formula, the height $h_{\overline{\mathcal{L}}_U}(P_U)$ of the point $P_U \in X_U$ is defined as follows. Consider the canonical integral model $\mathcal{X}_U$ of X_U over $\mathcal{O}_F$ and let $\overline{\mathcal{L}}_U$ be the arithmetic Hodge bundle of $\mathcal{X}_U$ involving the hermitian metric $\|dz\|_v = 2\mathrm{im}(z)$ at an archimedean place v of F. We refer to §4.2.6 for the constructions of $\mathcal{X}_U$ and $\overline{\mathcal{L}}_U$. Then the height $h_{\overline{\mathcal{L}}_U}(P_U)$ is defined by

$$h_{\overline{\mathcal{L}}_U}(P_U) = \tfrac{1}{[F(P_U):F]} \widehat{\deg}(\overline{\mathcal{L}}_U|_{\overline{P}_U}). \tag{4.13}$$

Here $F(P_U)$ is the field of definition of P_U, and $\overline{P}_U$ is the Zariski closure of $P_U \in X_U$ in $\mathcal{X}_U$. We remark that the left-hand side of (4.12) is precisely the averaged sum of the heights $h(\Phi)$ of CM types Φ of E appearing in the averaged Colmez conjecture (Theorem 4.14).

4.2.1 The Proof of Theorem 4.21

To give an overview of the strategy of proof, we continue to use the previous notation and we consider a point P on the quaternionic Shimura curve X as in Theorem 4.21. The proof can be divided into two parts, which then roughly speaking can be described as follows:

(i) To compute $\frac{1}{2^g} \sum_\Phi h(\Phi)$ in terms of $h_{\overline{\mathcal{L}}_U}(P_U)$, we decompose each Faltings height $h(\Phi)$ into smaller parts. This allows us to write $\frac{1}{2^g} \sum_\Phi h(\Phi)$ as a sum $\sum_{(\Phi_1, \Phi_2)} h(\Phi_1, \Phi_2)$ taken over certain pairs of CM types (Φ_1, Φ_2) of E, where $h(\Phi_1, \Phi_2)$ is essentially the height of a point P' on some PEL-type Shimura curve X' associated to (Φ_1, Φ_2).

(ii) To relate such points P' with our given point P, we construct another Shimura curve X'' together with morphisms $X \to X''$ and $X' \to X''$ such that P and P' have the same image in X''. Then we compute the height

$h(\Phi_1, \Phi_2)$ of P' in terms of the height $h_{\overline{\mathcal{L}}_U}(P_U)$ of P by constructing (first generically, then integrally) an isometry between two hermitian line bundles whose Arakelov degrees essentially determine these heights.

The proof of Theorem 4.21 combines various new ideas and concepts. In the following overview of §4.2, we briefly describe the main ideas and concepts and we provide an outline of the proof of Theorem 4.21 which describes in more detail the content of parts (i) and (ii).

Decomposition of the Faltings height. In part (i), we first decompose the Faltings height $h(\Phi)$ of any CM type Φ of E into its τ-parts $h(\Phi, \tau)$ with $\tau \in \Phi$. More generally, in §4.2.3 we discuss in detail the construction and various properties of the τ-part

$$h(A, \tau)$$

of the Faltings height of any abelian variety A defined over a number field such that A is equipped with an $\mathcal{O}_L$-action, where L is either a totally real number field or a CM field and where $\tau : L \hookrightarrow \mathbb{C}$ is an embedding. The τ-part $h(A, \tau)$ is defined as the normalized Arakelov degree of a hermitian line bundle $\overline{\mathcal{N}}(A, \tau)$, whose construction in (4.21) involves the τ-isotypic piece of the dual of the Lie algebra of the Néron model of A. The metric of $\overline{\mathcal{N}}(A, \tau)$ is defined via a natural pairing on the sheaf of differentials of a complex abelian variety and its dual.

Nearby pairs of CM types. Continuing in part (i), we crucially use the concept of nearby pairs of CM types (Φ_1, Φ_2) of E with $|\Phi_1 \cap \Phi_2| = g - 1$ to study the Faltings height of an abelian variety with complex multiplication. In §4.2.4, we begin by discussing basic properties of such pairs (Φ_1, Φ_2) and their height $h(\Phi_1, \Phi_2) = \frac{1}{2}\big(h(\Phi_1, \tau_1) + h(\Phi_2, \tau_2)\big)$, where $\tau_i \in \Phi_i - (\Phi_1 \cap \Phi_2)$ for $i \in \{1, 2\}$. Then, we use the decomposition of the Faltings height into τ-parts to obtain

$$\frac{1}{2^g} \sum_{\Phi} h(\Phi) = \frac{1}{2^{g-1}} \sum_{(\Phi_1, \Phi_2)} h(\Phi_1, \Phi_2) - \frac{1}{4} \log \Delta_F. \tag{4.14}$$

Here the sum on the right-hand side is taken over all non-ordered nearby pairs of CM types of E. We further explain that any nearby pair (Φ_1, Φ_2) of CM types of E satisfies

$$h(\Phi_1, \Phi_2) = \frac{1}{2} h(A_0, \tau) \tag{4.15}$$

where $\tau = \tau_1|_F$ and where A_0 is any abelian variety with $\mathcal{O}_E$-action such that A_0 is $\mathcal{O}_E$-isogenous to the product $A_{\Phi_1} \times A_{\Phi_2}$ of the CM abelian varieties A_{Φ_1} and A_{Φ_2} defined in (4.5). Here $h(A_0, \tau)$ involves the $\mathcal{O}_F$-action on A_0, and somewhat surprisingly $h(A_0, \tau)$ is independent of (Φ_1, Φ_2).

The Shimura curves X, X', and X''. In part (ii) we relate the image $P_U \in X_U = X/U$ of our given point $P \in X$ to some abelian variety A_0 as in (4.15) by working with several Shimura curves related to our curve X. To each nearby pair of CM types (Φ_1, Φ_2) of E, we associate in §4.2.5 a certain PEL-type

Shimura curve $X'_{U'}$ with a universal abelian scheme $\mathcal{A} \to X'_{U'}$. Moreover, in §4.2.7 we construct another Shimura curve $X''_{U'''}$ together with morphisms

$$X_U \to X''_{U'''}, \quad X'_{U'} \to X''_{U'''}. \tag{4.16}$$

For our given $P_U \in X_U$, we then find $P'_{U'} \in X'_{U'}$ such that $\mathcal{A}_{P'_{U'}} = A_0$ for some A_0 as in (4.15) and such that the maps in (4.16) send P_U and $P'_{U'}$ to the same point in $X''_{U'''}$. Also, in § 4.2.6 we define the arithmetic Hodge bundle $\overline{\mathcal{L}}_U$ of the canonical integral model $\mathcal{X}_U$ of X_U over $\mathcal{O}_F$ and we explain the construction of $\mathcal{X}_U$ which builds on the work of Carayol and Cerednik-Drinfeld.

Kodaira-Spencer isomorphisms. In part (ii) we compute $h(A_0, \tau)$ in terms of $h_{\overline{\mathcal{L}}_U}(P_U)$ for any A_0 as in (4.15). For this purpose, we use various Kodaira-Spencer isomorphisms in order to construct an isometry between the following two hermitian line bundles on $\mathrm{Spec}(\mathcal{O}_K)$:

$$\overline{\mathcal{N}}(A_0, \tau) \simeq (\overline{\mathcal{N}}_U|_{P_U}) \otimes_{\mathcal{O}_{F(P_U)}} \mathcal{O}_K. \tag{4.17}$$

Here $\overline{\mathcal{N}}_U = \overline{\mathcal{L}}_U^2(-\mathfrak{d}_{\mathbb{B}})$ is a $\mathbb{Q}$-line bundle over $\mathcal{X}_U$, where $\mathfrak{d}_{\mathbb{B}}$ denotes the product of the prime ideals of F over which $\mathbb{B}$ ramifies. Furthermore, $K \subseteq \mathbb{C}$ is a number field containing the normal closure of E such that any abelian variety with CM by $\mathcal{O}_E$ is defined over K and has good reduction over $\mathcal{O}_K$. We take the normalized arithmetic degree in (4.17) and then (4.15) gives

$$2gh(\Phi_1, \Phi_2) = h_{\overline{\mathcal{L}}_U}(P_U) - \tfrac{1}{2} \log \Delta_{\mathbb{B}} \tag{4.18}$$

for each nearby pair (Φ_1, Φ_2) of CM types of E. The isometry (4.17) is first constructed generically (over K) by using (4.16) and by applying an archimedean Kodaira-Spencer isomorphism involving the universal $\mathcal{A}$ over $X'_{U'}$; see §4.2.5. Then the isometry of line bundles over K is extended to $\mathcal{O}_K$ by working with p-divisible groups and by using a non-archimedean Kodaira-Spencer isomorphism; see §4.2.6. Finally, on inserting (4.18) into (4.14) we obtain

$$\frac{1}{2^g} \sum_{\Phi} h(\Phi) = \tfrac{1}{2} h_{\overline{\mathcal{L}}_U}(P_U) - \tfrac{1}{4} \log(\Delta_{\mathbb{B}} \Delta_F)$$

which is the formula claimed in Theorem 4.21. We mention that in §4.2.8 we shall give a detailed proof of Theorem 4.21 by combining the results collected in the following sections.

4.2.2 Notation and Terminology

We fix the following notation and terminology for the remainder of §4.2. Let E be either a CM field or a totally real number field. We denote by $c \colon E \to E$ the CM involution on E, where c is trivial if E is totally real. Let $\overline{\mathbb{Q}} \subseteq \mathbb{C}$ be an algebraic closure of $\mathbb{Q}$. We will identify the two sets of field morphisms: $\mathrm{Hom}(E,\mathbb{C}) = \mathrm{Hom}(E,\overline{\mathbb{Q}})$.

4.2.3 Decomposition of the Faltings Height

In this section, we first define the τ-part of the Faltings height of an abelian variety with $\mathcal{O}_E$-action, where τ is a complex embedding of E. For any CM type Φ of E, we then decompose the Faltings height of an abelian variety with CM by $(\mathcal{O}_E, \Phi)$ into its τ-parts with $\tau \in \Phi$.

Let A be a nonzero abelian variety defined over a number field K and denote by $\mathcal{A}$ the Néron model of A over $\mathcal{O}_K$ with unit section $e \colon \mathrm{Spec}(\mathcal{O}_K) \to \mathcal{A}$. We suppose that $K \subseteq \overline{\mathbb{Q}} \subseteq \mathbb{C}$ and that K contains the normal closure of E in $\overline{\mathbb{Q}}$. Then it holds in particular that $\mathrm{Hom}(E,\mathbb{C}) = \mathrm{Hom}(E,\overline{\mathbb{Q}}) = \mathrm{Hom}(E,K)$. Moreover, we assume that $\mathcal{A}$ is a semi-abelian scheme over $\mathcal{O}_K$ and that there exists a ring homomorphism $\iota \colon \mathcal{O}_E \to \mathrm{End}(A)$.

The τ-quotient space $\mathcal{W}(A,\tau)$. We consider the sheaf $\Omega(A) = \mathrm{Lie}(\mathcal{A})^\vee$ on $\mathrm{Spec}(\mathcal{O}_K)$, where $\mathrm{Lie}(\mathcal{A})^\vee$ denotes the dual of the Lie algebra $\mathrm{Lie}(\mathcal{A})$ associated to the functor of points of $\mathcal{A}$. There exists a canonical isomorphism (see, for example, [34, Prop. 1.1 (b)])

$$\Omega(A) = \mathrm{Lie}(\mathcal{A})^\vee \simeq e^* \Omega^1_{\mathcal{A}/\mathcal{O}_K}.$$

Here, the sheaf $\Omega^1_{\mathcal{A}/\mathcal{O}_K}$ of relative differential 1-forms of $\mathcal{A}/\mathcal{O}_K$ is a locally free sheaf of rank $\dim A$ since $\mathcal{A}$ is smooth over $\mathcal{O}_K$. The ring morphism $\iota \colon \mathcal{O}_E \to \mathrm{End}(A)$ induces an action of $\mathcal{O}_E \otimes \mathcal{O}_K$ on $\Omega(A)$. Indeed, on identifying $\mathrm{End}(A) \cong \mathrm{End}(\mathcal{A})$ via the Néron mapping property of $\mathcal{A}$, the morphism ι induces an action of $\mathcal{O}_E$ on $\mathcal{A}$. By functoriality of the Lie algebra, this then induces an $\mathcal{O}_K$-linear action of $\mathcal{O}_E$ on $\mathrm{Lie}(\mathcal{A})$ giving the action of $\mathcal{O}_E \otimes \mathcal{O}_K$ on $\Omega(A)$.

Let $\tau \in \mathrm{Hom}(E,K)$. On using the $\mathcal{O}_E$-action on A, we now construct a quotient of the sheaf $\Omega(A)$ in the following way. The embedding $\tau \colon E \to K$ induces a projection $\mathcal{O}_E \otimes \mathcal{O}_K \to \mathcal{O}_K$ which we continue to denote by τ. Now, on using the just-described $\mathcal{O}_E \otimes \mathcal{O}_K$-module structure on $\Omega(A)$, we define the τ-quotient $\mathcal{W}(A,\tau)$ of $\Omega(A)$ to be

$$\mathcal{W}(A,\tau) = \Omega(A) \otimes_{\mathcal{O}_E \otimes \mathcal{O}_K, \tau} \mathcal{O}_K.$$

Then $\mathcal{W}(A,\tau)$ is a locally free sheaf on $\mathrm{Spec}(\mathcal{O}_K)$. We remark that in the above construction of $\Omega(A)$ and $\mathcal{W}(A,\tau)$ one can replace the Néron model $\mathcal{A}$ by its identity component $\mathcal{A}^0$ since there exists a canonical isomorphism (see, for example, [34, Prop. 1.1 (d)])

$$\mathrm{Lie}(\mathcal{A}^0) \simeq \mathrm{Lie}(\mathcal{A}). \tag{4.19}$$

In particular, our $\Omega(A)$ and $\mathcal{W}(A,\tau)$ canonically identify with the corresponding locally free sheaves on $\mathrm{Spec}(\mathcal{O}_K)$ which were constructed in [60, §2] by using $\mathcal{A}^0$ in place of $\mathcal{A}$.

The τ-component $\mathcal{N}(A, \tau)$.   Let $\tau\colon E \to K$ be an embedding. We now explain how to endow the line bundle $\det \mathcal{W}(A, \tau)$ on $\mathrm{Spec}(\mathcal{O}_K)$ with a metric which is compatible with the canonical hermitian metric (4.1) of the Hodge bundle $\omega(\mathcal{A}) = \det \Omega(\mathcal{A})$ used in the definition of the Faltings height. To avoid making a choice of polarization of A, we work simultaneously with the dual of A and consider the line bundle $\mathcal{N}(A, \tau)$ on $\mathrm{Spec}(\mathcal{O}_K)$ defined as

$$\mathcal{N}(A, \tau) = \det \mathcal{W}(A, \tau) \otimes \det \mathcal{W}(A^t, \tau^c).$$

Here $\tau^c = \tau \circ c$ is the composition of τ with the CM involution $c\colon E \to E$, and $\mathcal{W}(A^t, \tau^c)$ is a locally free sheaf on $\mathrm{Spec}(\mathcal{O}_K)$ whose construction we now explain in detail. Let $A^t = \mathrm{Pic}^0(A)$ denote the dual of A. Then A^t is an abelian variety defined over K of dimension $\dim A$. Its Néron model over $\mathcal{O}_K$, which we denote by $\mathcal{A}^t$, is a semi-abelian scheme. Further, the functoriality of Pic^0 assures that sending $x \in \mathcal{O}_E$ to the endomorphism $\mathrm{Pic}^0(\iota(x))$ of A^t defines a ring morphism

$$\iota^\vee\colon \mathcal{O}_E \to \mathrm{End}(A^t).$$

In exactly the same way as in the construction of $\mathcal{W}(A, \tau)$, we see that $\iota^\vee$ induces an $\mathcal{O}_E \otimes \mathcal{O}_K$ action on $\Omega(A^t)$. Then we define the locally free sheaf $\mathcal{W}(A^t, \tau)$ on $\mathrm{Spec}(\mathcal{O}_K)$ by

$$\mathcal{W}(A^t, \tau) - \Omega(A^t) \otimes_{\mathcal{O}_E \otimes \mathcal{O}_K, \tau} \mathcal{O}_K.$$

Here we used again the projection $\mathcal{O}_E \otimes \mathcal{O}_K \to \mathcal{O}_K$ induced by our embedding $\tau\colon E \to K$.

The metric on the τ-component $\mathcal{N}(A, \tau)$.   We take again an embedding $\tau\colon E \to K$. To metrize the line bundle $\mathcal{N}(A, \tau)$ on $\mathrm{Spec}(\mathcal{O}_K)$, we use a hermitian pairing on the sheaf of invariant differential 1-forms of a complex abelian variety and its dual. Let v be an archimedean (infinite) place of K. In what follows in this subsection we denote by the subscript v the base change from K to $\mathbb{C}$ via an embedding $K \hookrightarrow \mathbb{C}$ associated to v. First, we define

$$\Omega(A_v) = \Omega(A) \otimes_v \mathbb{C} \quad \text{and} \quad \Omega(A_v^t) = \Omega(A^t) \otimes_v \mathbb{C}.$$

It holds that $\Omega(A_v) \simeq H^0(A_v, \Omega^1_{A_v/\mathbb{C}})$ and $\Omega(A_v^t) \simeq H^0(A_v^t, \Omega^1_{A_v^t/\mathbb{C}})$. To construct the τ-quotient spaces of $\Omega(A_v)$ and $\Omega(A_v^t)$, we use that $\tau\colon E \to K \subseteq \mathbb{C}$ defines a projection $E \otimes \mathbb{C} \to \mathbb{C}$ which we continue to denote by τ. Further, as in the above-explained construction of $\mathcal{W}(A, \tau)$ and $\mathcal{W}(A^t, \tau)$, we see that the ring morphisms ι and $\iota^\vee$ induce an action of $E \otimes \mathbb{C}$ on $\Omega(A_v) \simeq \mathrm{Lie}(A_v)^\vee$ and $\Omega(A_v^t) \simeq \mathrm{Lie}(A_v^t)^\vee$, respectively. Then we define the τ-quotient spaces

$$W(A_v, \tau) = \Omega(A_v) \otimes_{E \otimes \mathbb{C}, \tau} \mathbb{C} \quad \text{and} \quad W(A_v^t, \tau) = \Omega(A_v^t) \otimes_{E \otimes \mathbb{C}, \tau} \mathbb{C}.$$

The modules $\Omega(A_v)$ and $\Omega(A_v^t)$ thus admit the decompositions

$$\Omega(A_v) = \bigoplus_{\tau:\, E \to K} W(A_v, \tau) \quad \text{and} \quad \Omega(A_v^t) = \bigoplus_{\tau:\, E \to K} W(A_v^t, \tau)$$

into the τ-quotient spaces $W(A_v, \tau)$ and $W(A_v^t, \tau)$, respectively. Both direct sums above are taken over all embeddings $\tau \colon E \to K \subseteq \mathbb{C}$. Furthermore, there is a perfect hermitian pairing

$$(\cdot, \cdot)_{A_v} : \Omega(A_v) \times \Omega(A_v^t) \longrightarrow \mathbb{C}.$$

This pairing is induced from the complex uniformization $A_v^t(\mathbb{C}) \simeq \mathrm{H}^1(A_v, \mathcal{O}_{A_v})/\mathrm{H}^1(A_v, 2\pi i \mathbb{Z})$ together with the canonical isomorphisms

$$\Omega(A_v^t)^\vee \simeq \mathrm{H}^1(A_v, \mathcal{O}_{A_v}) \simeq \mathrm{H}^{0,1}(A_v) \simeq \overline{\Omega(A_v)}.$$

The pairing $(\cdot, \cdot)_{A_v}$ is functorial in the following sense: For any morphism of complex abelian varieties $\phi \colon B \to C$ with dual $\phi^t \colon C^t \to B^t$, it holds that $(\phi^* \gamma, \beta)_B = (\gamma, (\phi^t)^* \beta)_C$ for all $\gamma \in \mathrm{Lie}(C)^\vee$ and $\beta \in \mathrm{Lie}(B^t)^\vee$. Moreover, the pairing $(\cdot, \cdot)_{A_v}$ has the following properties:

1. Taking determinants, $(\cdot, \cdot)_{A_v}$ induces a hermitian norm $\| \cdot \|_v$ on $\omega(A_v) \otimes \omega(A_v^t)$, where $\omega = \det \Omega$. Then for all $\alpha \in \omega(A_v)$ and $\beta \in \omega(A_v^t)$, it holds (see [60, Lem 2.1]):
$$\|\alpha\|_v \cdot \|\beta\|_v = \|\alpha \otimes \beta\|_v.$$

   Here both norms appearing on the left-hand side are the canonical hermitian norms (4.1) used in the definition of the Faltings height.

2. The hermitian pairing $(\cdot, \cdot)_{A_v}$ is the orthogonal sum of hermitian pairings

$$W(A_v, \tau) \times W(A_v^t, \tau^c) \longrightarrow \mathbb{C} \tag{4.20}$$

   with the sum taken over all embeddings $\tau \colon E \to K \subseteq \mathbb{C}$.

We are now ready to define the metric $\| \cdot \|_\tau$ on the line bundle $\mathcal{N}(A, \tau)$ on $\mathrm{Spec}(\mathcal{O}_K)$: For each archimedean place v of K, we endow the line bundle

$$\mathcal{N}(A, \tau)_v \simeq \det W(A_v, \tau) \otimes \det W(A_v^t, \tau^c)$$

on $\mathrm{Spec}(\mathbb{C})$ with the metric defined by the hermitian pairing (4.20). Then we define

$$\overline{\mathcal{N}}(A, \tau) = (\mathcal{N}(A, \tau), \| \cdot \|_\tau). \tag{4.21}$$

This is a hermitian line bundle on $\mathrm{Spec}(\mathcal{O}_K)$.

The τ-part of the Faltings height.   Let $\tau\colon E\to K$ be an embedding. Using the hermitian line bundle $\overline{\mathcal{N}}(A,\tau)$ on $\mathrm{Spec}(\mathcal{O}_K)$, we define the τ-part $h(A,\tau)$ of the Faltings height $h(A)$ by

$$h(A,\tau)=\frac{1}{2[K:\mathbb{Q}]}\widehat{\deg}(\overline{\mathcal{N}}(A,\tau)). \tag{4.22}$$

Here the factor $\frac12$ takes into account that we worked simultaneously with the dual. We point out that $h(A,\tau)$ a priori depends on our initial choice of the ring morphism $\iota\colon\mathcal{O}_E\to\mathrm{End}(A)$. To emphasize this dependence, we shall write $h(A,\iota,\tau)$ for $h(A,\tau)$ in situations where the choice of the involved ι is important. Further, $h(A,\tau)$ depends a priori on K since we assumed that $\mathcal{A}$ is semi-abelian over $\mathcal{O}_K$ and that K contains the normal closure of E in $\overline{\mathbb{Q}}$. In the following paragraph we give a construction which will turn out to be independent of K.

The stable τ-part of the Faltings height.   Let $K\subseteq\mathbb{C}$ be a number field and let A be a nonzero abelian variety over K together with a ring morphism $\iota\colon\mathcal{O}_E\to\mathrm{End}(A)$. There exists a finite field extension L/K with $L\subseteq\mathbb{C}$ such that the Néron model of $A_L=A\times_K L$ over $\mathcal{O}_L$ is semi-abelian and such that L contains the normal closure of E in $\mathbb{C}$. For any embedding $\tau\colon E\to\mathbb{C}$, we define the stable τ-part $h(A,\tau)$ of the Faltings height $h(A)$ by

$$h(A,\tau)=h(A_L,\iota,\tau). \tag{4.23}$$

This is independent of the choice of any field extension L/K as above since the identity component of a semi-abelian Néron model commutes with base change. However, the stable τ-part $h(A,\tau)$ still depends a priori on the choice of the ring morphism $\iota\colon\mathcal{O}_E\to\mathrm{End}(A)$. Thus we shall write $h(A,\iota,\tau)$ for $h(A,\tau)$ in situations where the choice of the involved ι is important.

We emphasize that for everything we mentioned so far in this §4.2.3, the field E can be either a totally real field or a CM field, and we do not require conditions on $[E:\mathbb{Q}]$ or $\dim A$. The following holds in the case *when E is CM* of degree $[E:\mathbb{Q}]=2\dim A$. In this setting, the existence of ι assures that A has CM by $(\mathcal{O}_E,\Phi)$ for some CM type Φ of E and [60, Thm. 2.2] gives that $h(A,\tau)$ depends only on (Φ,τ). In light of this observation, we will use the following notation: For any CM field E, each CM type Φ of E and any embedding $\tau\colon E\to\mathbb{C}$, we write

$$h(\Phi,\tau)=h(A_\Phi,\tau). \tag{4.24}$$

Here $h(A_\Phi,\tau)$ denotes the above-constructed stable τ-part of the Faltings height $h(A_\Phi)$ of the CM abelian variety A_Φ defined in (4.5).

The decomposition of the Faltings height of Φ.   From now on we suppose that E is a CM field. Let Φ be a CM type of E and denote by $h(\Phi)$ its Faltings height defined in (4.6). To decompose $h(\Phi)$ into the stable τ-parts $h(\Phi, \tau)$ constructed above for all $\tau \in \Phi$, we denote by $E_\Phi = \mathbb{Q}(\sum_{\tau \in \Phi} \tau(x) : x \in E)$ the reflex field of Φ. Then [60, Thm. 2.3] gives

$$h(\Phi) - \sum_{\tau \in \Phi} h(\Phi, \tau) = -\tfrac{1}{4[E_\Phi : \mathbb{Q}]} \log(d_\Phi d_{\Phi^c}). \tag{4.25}$$

Here d_Φ and d_{Φ^c} are the discriminants of Φ and Φ^c, respectively, defined in [60, §2.2]. The proof of [60, Thm. 2.3] additionally shows that the formula (4.25) holds more generally with Φ replaced by any complex abelian variety A having CM by $(\mathcal{O}_E, \Phi)$. Further, we recall that the stable τ-part $h(A, \tau)$ of such an A depends only on (Φ, τ). Therefore, we obtain a new proof of a result of Colmez (which is a consequence of [13, Thm. 0.3 (ii)]) asserting that the Faltings height of any complex abelian variety with CM by $(\mathcal{O}_E, \Phi)$ depends only on Φ.

4.2.4 The Faltings Height of Nearby Pairs of CM Types

In this section we first define a height $h(\Phi_1, \Phi_2)$ of a *nearby pair of CM types* (Φ_1, Φ_2) of a CM field E. We then use the decomposition of the Faltings height in (4.25) to compute the averaged sum $\frac{1}{2^g} \sum_\Phi h(\Phi)$ of the heights of the CM types Φ of E in terms of the heights $h(\Phi_1, \Phi_2)$. Finally, we express each $h(\Phi_1, \Phi_2)$ as the τ-part of the Faltings height of certain explicitly constructed abelian varieties A_0 which define points on the Shimura curves considered in the next section.

From now on we specialize to the case when E is a CM field of degree $[E : \mathbb{Q}] = 2g$ for some integer $g \geq 1$. We denote by F the maximal totally real subfield of E.

Nearby pairs of CM types. After defining nearby pairs of CM types of E, we construct their height by using the stable τ-part of the Faltings height. Let Φ_1 and Φ_2 be two CM types of E. The pair (Φ_1, Φ_2) of CM types of E is called *nearby* if

$$|\Phi_1 \cap \Phi_2| = g - 1.$$

We now assume that (Φ_1, Φ_2) is a pair of nearby CM types of E, and we denote by $\tau_i : E \to \mathbb{C}$ the embedding in the complement of $\Phi_1 \cap \Phi_2$ in Φ_i for $i = 1, 2$. Then we define

$$h(\Phi_1, \Phi_2) = \frac{1}{2}\left(h(\Phi_1, \tau_1) + h(\Phi_2, \tau_2)\right).$$

Here $h(\Phi_i, \tau_i)$ is the stable τ_i-part of the Faltings height of Φ_i defined in (4.24). The restrictions of the two embeddings $\tau_1, \tau_2 : E \to \mathbb{C}$ to the maximal totally real subfield $F \subseteq E$ coincide. Indeed, it holds that $\tau_1 = \tau_2 \circ c$, since $\Phi_i \cup \Phi_i^c = \mathrm{Hom}(E, \mathbb{C})$ and $|\Phi_1 \cap \Phi_2| = g - 1$. We now use the heights $h(\Phi_1, \Phi_2)$, to compute the sum $\sum_\Phi h(\Phi)$ of the heights of CM types Φ of E.

Computing $\sum_\Phi h(\Phi)$ in terms of $h(\Phi_1, \Phi_2)$. We recall that $[E:\mathbb{Q}] = 2g$ and that Δ_F denotes the absolute discriminant of the totally real number field F. Then [60, Cor. 2.6] gives

$$\frac{1}{2^g} \sum_\Phi h(\Phi) = \frac{1}{2^{g-1}} \sum_{(\Phi_1, \Phi_2)} h(\Phi_1, \Phi_2) - \frac{1}{4} \log \Delta_F. \qquad (4.26)$$

The sum on the left is taken over all CM types Φ of E, and the sum on the right runs over all non-ordered nearby pairs (Φ_1, Φ_2) of CM types of E. Also, we recall that $h(\Phi)$ denotes the Faltings height of Φ defined in (4.6). We mention that the proof of the formula (4.26) given in [60] uses the decomposition of the Faltings height $h(\Phi)$ obtained in (4.25).

Writing $h(\Phi_1, \Phi_2)$ as $\frac{1}{2}h(A_0, \tau)$ for some A_0 and τ. Let (Φ_1, Φ_2) be a nearby pair of CM types of E. We show that $2h(\Phi_1, \Phi_2)$ can be written as the stable τ-part of the Faltings height of certain abelian varieties. For $i \in \{0, 1, 2\}$, let A_i be an abelian variety over $\overline{\mathbb{Q}}$ endowed with a ring morphism $\iota_i : \mathcal{O}_E \to \mathrm{End}(A_i)$. We make the following assumptions:

(i) the abelian variety A_i has CM by $(\mathcal{O}_E, \Phi_i)$ with Φ_i induced by ι_i for $i \in \{1, 2\}$;

(ii) the abelian varieties A_0 and $A_1 \times A_2$ are $\mathcal{O}_E$-isogenous.

Assumption (ii) means that there is an isogeny $f : A_0 \to A_1 \times A_2$ such that $f\iota_0(x) = \iota'(x)f$ for all $x \in \mathcal{O}_E$, where ι' is the ring morphism $\mathcal{O}_E \to \mathrm{End}(A_1 \times A_2)$ given by the product $\iota_1 \times \iota_2$.

Let $\tau : F \to \mathbb{C}$ be the embedding defined by the restriction $\tau = \tau_1|_F = \tau_2|_F$, where $\tau_i : E \to \mathbb{C}$ lies in the complement of $\Phi_1 \cap \Phi_2$ in Φ_i for $i \in \{1, 2\}$. Now, given abelian varieties A_0, A_1, A_2 as above which satisfy (i) and (ii), the proof of [60, Thm. 2.7] gives

$$h(\Phi_1, \Phi_2) = \tfrac{1}{2} h(A_0, \iota_0|_{\mathcal{O}_F}, \tau). \qquad (4.27)$$

Here $h(A_0, \iota_0|_{\mathcal{O}_F}, \tau)$ is the stable τ-part (4.23) of the Faltings height of A_0 together with the ring morphism $\iota_0|_{\mathcal{O}_F} : \mathcal{O}_F \to \mathrm{End}(A_0)$ obtained by restricting ι_0 to the subring $\mathcal{O}_F \subseteq \mathcal{O}_E$. In view of the formulas (4.26) and (4.27), we can compute the averaged sum $\frac{1}{2^g} \sum_\Phi h(\Phi)$ if for each nearby pair of CM types (Φ_1, Φ_2) of E we can construct A_0, A_1, A_2 and if we can then compute the stable τ-part $h(A_0, \iota_0|_{\mathcal{O}_F}, \tau)$. We next discuss the construction of the abelian varieties A_0, A_1, A_2.

The construction of A_0. Let (Φ_1, Φ_2) be a nearby pair of CM types of E, and for $i \in \{1, 2\}$ let A_{Φ_i} be the CM abelian variety associated to Φ_i in (4.5). Then we observe that $A_1 = A_{\Phi_1}$ and $A_2 = A_{\Phi_2}$ satisfy the above condition (i). We now construct an abelian variety A_0 satisfying the above condition (ii). Let B/F be a quaternion algebra and assume that B contains E. Then there exists $j \in B$ such that $B = E + Ej$ with $jx = \bar{x}j$ for all $x \in E$. Let $\mathcal{O}_B \supseteq \mathcal{O}_E$ be a maximal order of the quaternion algebra B. There exists a complex abelian variety A_0 with

$$A_0(\mathbb{C}) \simeq \frac{B \otimes \mathbb{R}}{\mathcal{O}_B} \tag{4.28}$$

where the complex structure on $B \otimes \mathbb{R}$ is given by the isomorphism

$$B \otimes \mathbb{R} = (E \otimes \mathbb{R}) \oplus (E \otimes \mathbb{R}) \xrightarrow[\sim]{(\Phi_1, \Phi_2)} \mathbb{C}^g \oplus \mathbb{C}^g$$

for which the first identification is induced by $B = E + Ej$. By the complex structure, A_0 is isogenous to $A_1 \times A_2$. Thus, the abelian variety A_0 has CM by $(E, \Phi_1 + \Phi_2)$ and therefore we may and do assume that A_0 is defined over $\overline{\mathbb{Q}}$. Moreover, on restricting the natural $\mathcal{O}_B$-action on A_0 to the subring $\mathcal{O}_E \subseteq \mathcal{O}_B$, we obtain a ring morphism $\iota_0 : \mathcal{O}_E \to \text{End}(A_0)$. It turns out that the natural isogeny $A_0 \to A_1 \times A_2$ can be taken to be an $\mathcal{O}_E$-isogeny, and so we conclude that A_0 satisfies assumption (ii). In other words, A_0, A_1, A_2 have all the desired properties.

4.2.5　A Shimura Curve of PEL-type $(\Phi_1 + \Phi_2)$

The various moduli problems. We will study in this section a Shimura curve of PEL-type which parametrizes the abelian varieties $A_0 := \frac{B \otimes \mathbb{R}}{\mathcal{O}_B}$ introduced in the end of §4.2.4. We continue to denote by E a CM field such that $[E : \mathbb{Q}] = 2g$ and by $F \subseteq E$ its maximal totally real subfield.

Let Σ denote the set of archimedean places of F. We impose the condition that Σ has odd cardinality. Fix $\tau \in \Sigma$ and let B/F be a quaternion algebra which is split at τ and otherwise ramified at every element in $\Sigma \setminus \{\tau\}$.

For technical reasons, we consider a situation slightly more general than the one presented in §4.2.4. Assume that we are in one of the following two cases:

(i) $E = F(\sqrt{\lambda})$ for some $\lambda \in \mathbb{Q}$.

(ii) $E \subseteq B$, so there exists $j \in B$ such that $B = E + Ej$ with $jx = \bar{x}j$ for all $x \in E$.

Case (ii), being the situation we considered in §4.2.4, is the main case. The Shimura curve of PEL-type X' parametrizing A_0 is constructed under this case. Hence, in order to reduce the computation of $h(A_0, \tau)$ to the right-hand side of Theorem 4.21, we need to understand the relation between X' and the Shimura curve X of quaternion type (see §4.2.6). This relation will be explained in §4.2.7.

Case (i) is needed in the study of an integral model of X when computing the height of its rational points. Obtaining such a good integral model uses Carayol's work [7], which relates X with a Shimura curve of PEL-type under case (i).

The reader may ask at this point why we do not simply compute everything with X' under case (ii). Here are the reasons. First, we need to vary the nearby CM pair (Φ_1, Φ_2). For each (Φ_1, Φ_2) we obtain a different X', whereas X is defined by a fixed quaternion algebra and as such is independent of the nearby CM pair. Second, there is no natural extension of the moduli problem parametrized by X' under case (ii) to the ring of integers (see the end of this subsection for a precise statement). In the paper of Yuan-Zhang [60], the integral model of this X' is constructed via the comparison between X' and X. Third, to compute the right-hand side of Theorem 4.21, i.e., the height of the special point on X, one requires deep knowledge of automorphic forms of $B^\times$. By working with X' we would still have to recover this computation, so it is much more convenient to work directly with the Shimura curve of quaternion type X.

Constructing the PEL Shimura varieties of type $\Phi_1 + \Phi_2$. Let us define the Shimura curves X' in both cases above. In either case, we let $B^1 \subseteq B$ and $E^1 \subseteq E$ be the subgroups of norm 1 elements. Let Φ_1, Φ_2 be nearby CM types of E, such that $\tau_i|_F = \tau$, where $\tau_i \in \Phi_1 \setminus (\Phi_1 \cap \Phi_2)$ for $i \in \{1, 2\}$. Define the reflex field F' of $\Phi_1 + \Phi_2$ to be the field of $\Phi_1 + \Phi_2$-traces:

$$F' := \mathbb{Q}(\mathrm{tr}_{\Phi_1 + \Phi_2}(t) : t \in E). \tag{4.29}$$

Then $F' \supseteq \tau(F)$ (see [60, Prop. 3.1]). By the linearity of the Φ-trace, $\mathrm{tr}_{\Phi_1 + \Phi_2}(t) = \mathrm{tr}_{\Phi_1}(t) + \mathrm{tr}_{\Phi_2}(t)$.

Define the quaternion algebra $B' := B \otimes_F E$ over E and let $V' := B'$ viewed as a $\mathbb{Q}$-vector space with a left action of B' given by multiplication in B'. Denote by $b \mapsto \bar{b}$ the involution on B' induced from the canonical involution on B and the complex conjugation on E. After fixing an invertible element $\gamma' \in B'$ with $\overline{\gamma'} = -\gamma'$, we may also define a symplectic space $V' = (B', \psi')$ with the symplectic form ψ' given by

$$\psi'(u, v) = \mathrm{tr}_{E/\mathbb{Q}} \mathrm{tr}_{B'/E}(\gamma' u \bar{v}), \quad u, v \in V'. \tag{4.30}$$

Note that the inside trace in this formula is the *reduced* trace on the quaternion algebra. Define another involution $*: B' \to B'$ by $\ell \mapsto \gamma'^{-1} \bar{\ell} \gamma'$. Then $*$ satisfies $\psi'(\ell v, w) = \psi'(v, \ell^* w)$ for any $v, w \in V'$.

Let now G' be the group of B'-linear symplectic similitudes of (V', ψ'). This is an algebraic group whose L-points for any number field L is given concretely as follows:

$$G'(L) := \{g \in \mathrm{Aut}_{B' \otimes L}(V' \otimes L) : \exists r \in L^\times \text{ such that } \psi'(gu, gv)$$
$$= r \cdot \psi'(u, v) \text{ for all } u, v \in V' \otimes L\}.$$

The group G' also has the following description, which will be useful later to settle questions about constructing integral models. Abusing notation, since every reductive group we consider is defined over $\mathbb{Q}$, we will write a reductive group by its group of $\mathbb{Q}$-points (so, for example, we write $\mathrm{Res}_{F/\mathbb{Q}}(B^\times)$ by $B^\times$ and $\mathrm{Res}_{E/\mathbb{Q}}(\mathbb{G}_{m,E})$ by $E^\times$). Then define the reductive group $G'' := B^\times \times_{F^\times} E^\times$ over $\mathbb{Q}$ to be the quotient of $B^\times \times E^\times$ by $F^\times$ under the action $a \circ (b, e) = (ba^{-1}, ae)$, where $a \in F^\times$, $b \in B^\times$, and $e \in E^\times$. The derived subgroup of G'' is identified with B^1, and taking quotient of G'' by B^1 gives it an identification with $F^\times \times E^1$ via the map

$$\nu = (\nu_1, \nu_2) \colon G''/B^1 \to F^\times \times E^1$$

$$(b, e) \mapsto (\mathrm{tr}_{B/F}(b)e\bar{e}, e/\bar{e}).$$

It is a simple check that G' is given by the Grothendieck functor-of-points formalism as

$$G'(L) := \{g \in G''(L) : \nu_1(g) \in L^\times\}. \tag{4.31}$$

We wish to endow $V'(\mathbb{R})$ with a complex structure. Recall that for an $\mathbb{R}$-vector space, it is equivalent to endow it with a complex structure and with a Hodge structure of type $(-1, 0) + (0, -1)$. Let $\mathbb{S} = \mathrm{Res}_{\mathbb{C}/\mathbb{R}}(\mathbb{G}_{m,\mathbb{C}})$ be the Deligne torus. By definition G' is a subgroup of $\mathrm{GL}(V')$. Thus it suffices to find a group homomorphism $h' \colon \mathbb{S} \to G'_{\mathbb{R}}$ which induces the desired Hodge type on $V'(\mathbb{R})$.

The morphism h' is constructed in the following way. First, define the morphisms

$$h \colon \mathbb{S}(\mathbb{R}) = \mathbb{C}^\times \to B^\times(\mathbb{R}) \xrightarrow[\sim]{(\tau, \iota_2, \dots, \iota_g)} \mathrm{GL}_2(\mathbb{R}) \times (\mathbb{H}^\times)^{g-1} \tag{4.32}$$

$$z = x + iy \mapsto \left(\begin{pmatrix} x & y \\ y & -x \end{pmatrix}^{-1}, 1, \dots, 1 \right)$$

and

$$h_E \colon \mathbb{S}(\mathbb{R}) = \mathbb{C}^\times \to E^\times(\mathbb{R}) \xrightarrow[\sim]{\Phi_1} (\mathbb{C}^\times)^g \tag{4.33}$$

$$z \mapsto (1, z^{-1}, \dots, z^{-1}).$$

Here, $\{\tau, \iota_2, \dots, \iota_g\}$ are the archimedean places of F, and the first component of the map Φ_1 in (4.33) is τ_1. Then we have a morphism $\mathbb{S} \xrightarrow{(h, h_E)} B_{\mathbb{R}}^\times \times E_{\mathbb{R}}^\times \to G''_{\mathbb{R}}$. It is not hard to show that this morphism factors through $G'_{\mathbb{R}}$. This is the $h' \colon \mathbb{S} \to G'_{\mathbb{R}}$ we desire.

The $G'(\mathbb{R})$-conjugacy class of h' is identified with $\mathfrak{h}^\pm = \mathbb{C} \setminus \mathbb{R}$ by

$$gh'g^{-1} \mapsto g(i), \quad g \in G'(\mathbb{R}).$$

The pair $(G', \mathfrak{h}^\pm)$ satisfies the axioms of a Shimura datum, which in turn allows us to construct Shimura curves over $\mathbb{C}$ indexed by open and compact subgroups

$U' \subseteq G'(\widehat{\mathbb{Q}})$ as follows:

$$X'_{U'}(\mathbb{C}) := G'(\mathbb{Q}) \backslash \mathfrak{h}^{\pm} \times G'(\widehat{\mathbb{Q}})/U'. \tag{4.34}$$

Since the reflex field of h' is the same as the reflex field of $\Phi_1 + \Phi_2$, the Shimura curve $X'_{U'}$ has an algebraic model over F'. Moreover, Carayol [7, §3.1] demonstrated that for each infinite place τ' of F', the complex points $X'_{U',\tau'}(\mathbb{C})$ equals the right-hand side of (4.34).

For U' sufficiently small, the Shimura curve $X'_{U'}/F'$ solves the following moduli problem. For any F'-scheme S, $X'_{U'}(S)$ parametrizes equivalence classes of quadruples $(A, \lambda, \iota, \kappa)$, where

(1) A/S is an abelian scheme of relative dimension $4g$ up to isogeny,
(2) $\iota \colon B' \hookrightarrow \mathrm{End}^0(A/S)$ is a homomorphism such that the induced action of $\ell \in B'$ on the S-module $\mathrm{Lie}(A/S)$ has trace given by

$$\mathrm{tr}(\iota(\ell), \mathrm{Lie}(A/S)) = \mathrm{tr}_{\Phi_1 + \Phi_2}(\mathrm{tr}_{B'/E}(\ell)),$$

(3) $\lambda \colon A \to A^t$ is a polarization of A whose Rosati involution on $\mathrm{End}^0(A/S)$ induces the involution $*$ on B', and
(4) $\kappa \colon \widehat{V'} \times S \xrightarrow{\sim} H_1(A, \widehat{\mathbb{Q}})$ is a U'-orbit of similitudes of skew hermitian B'-modules, where $\widehat{V'} = V' \otimes_{\mathbb{Q}} \widehat{\mathbb{Q}}$.

Special case of the PEL Shimura variety of type $\Phi_1 + \Phi_2$. Now we focus on case (ii), i.e., when $B = E + Ej$ for some $j \in B$ with $jx = \overline{x}j$ for all $x \in E$. Thus we retain the setting in §4.2.4. Note that $\overline{x} = jxj^{-1}$ implies that $x = j^2 x j^{-2}$ for all $x \in E$. Hence $j^2 \in E$.

The previous discussion can be simplified in this case as follows. We can identify $B' = B \otimes_F E$ with $M_2(E)$ by

$$a \otimes b \mapsto \begin{pmatrix} ab & \\ & \overline{a}b \end{pmatrix}, \quad j \otimes 1 \mapsto \begin{pmatrix} & 1 \\ j^2 & \end{pmatrix}, \quad \forall a, b \in E.$$

We see $V' = B \otimes_F E$ as an E-vector space with the action of $e' \in E$ given by $e'(b \otimes e) = (e'b) \otimes e$. Let $V := B$ viewed as an E-vector space. Then we have an isomorphism of E-vector spaces

$$V' = B \otimes_F E \xrightarrow{\sim} V \oplus V, \quad b \otimes e \mapsto (eb, \overline{e}b).$$

The operator $w = \begin{pmatrix} & 1 \\ 1 & \end{pmatrix}$ switches the two factors by $(u, v) \mapsto (jv, j^{-1}u)$. In the rest of this discussion, we shall view an E-vector space as a $\mathbb{Q}$-vector space with an action of E on the left.

Next we decompose the symplectic form ψ' on V'. Take a purely imaginary element $\gamma \in E$, i.e., $\overline{\gamma} = -\gamma$, and let $\gamma' = \gamma \otimes 1 \in B \otimes_F E = B'$. Then $\overline{\gamma'} = -\gamma'$ and the symplectic form ψ' defined by (4.30) is the sum of two copies of the following symplectic form ψ on V defined by

$$\psi(u,v) = \mathrm{tr}_{F/\mathbb{Q}}\mathrm{tr}_{B/F}(\gamma u\bar{v}), \quad u,v \in V. \tag{4.35}$$

The inside trace in this formula is the *reduced* trace on the quaternion algebra. The group G' is identified with the group of E-linear symplectic similitudes of (V,ψ).

Now let $\mathcal{O}_B$ be a maximal order of B which contains $\mathcal{O}_E$, and let $\Lambda := \mathcal{O}_B$ be viewed as a lattice in V. Assume that ψ takes integral values on Λ. Define $\widehat{\Lambda} = \Lambda \otimes_{\mathbb{Z}} \widehat{\mathbb{Z}}$. Then for U' sufficiently small, the moduli problem solved by $X'_{U'}/F'$ has a simpler description. For any F'-scheme S, $X'_{U'}(S)$ parametrizes equivalence classes of quadruples (A,λ,ι,κ), where

(1) A/S is an abelian scheme of relative dimension $2g$,
(2) $\iota: \mathcal{O}_E \hookrightarrow \mathrm{End}(A/S)$ is a homomorphism such that the induced action of $\ell \in \mathcal{O}_E$ on the $\mathcal{O}_S$-module $\mathrm{Lie}(A/S)$ has trace given by

$$\mathrm{tr}(\iota(\ell), \mathrm{Lie}(A/S)) = \mathrm{tr}_{\Phi_1+\Phi_2}(\ell),$$

(3) $\lambda: A \to A^t$ is a polarization of A whose Rosati involution on $\mathrm{End}(A/S)$ induces the complex conjugation c of $\mathcal{O}_E$, and
(4) $\kappa: \widehat{\Lambda} \times S \to H_1(A,\widehat{\mathbb{Z}})$ is a U'-orbit of similitudes of skew hermitian $\mathcal{O}_E$-modules.

Remark 4.22. The reader may check that the h' obtained from (4.32) and (4.33) induces the complex structure $(\Phi_1,\Phi_2): V(\mathbb{R}) \xrightarrow{\sim} \mathbb{C}^g \times \mathbb{C}^g$. Hence the abelian variety A_0 constructed in §4.2.4 defines a point $[A_0] \in X'_{U'}(\mathbb{C})$ for the maximal

$$U' = \{g \in \mathrm{Aut}_{\widehat{\mathcal{O}}_E}(\widehat{\Lambda}): \psi(gu,gu) = \nu_1(g) \cdot \psi'(u,v) \text{ for all } u,v \in \widehat{\Lambda}\}$$

in the following way. If we denote by T' the subgroup of G' defined by $T'(\mathbb{Q}) = \{(b,e) \in G''(\mathbb{Q}) = B^\times \times_F E^\times : b \in E^\times\}$, then T' is a torus and $h' \in \mathfrak{h}^\pm$ is (the unique point) fixed by $T'(\mathbb{R})$. Let $(X'_{U'})^{T'}$ be the subscheme of $X'_{U'}$ of points fixed by T'. Each $\mathbb{C}$-point P' in $(X'_{U'})^{T'}$ is of the form $[h',t]$ for some $t \in T'(\widehat{\mathbb{Q}})$ under the expression of double coset (4.34), and hence represents a CM abelian variety $A_{P'}$ isogenous to $A_{\Phi_1} \times A_{\Phi_2}$. The point $[A_0]$ is one of them.

For a sufficiently small open compact subgroup $U' \subseteq G'(\widehat{\mathbb{Q}})$, this moduli problem is representable by a universal abelian variety $\pi: A_{U'} \to X'_{U'}$ on which $\mathcal{O}_E$ acts via $\Phi_1 + \Phi_2$. The reader may check that this abelian scheme is generically simple, and when restricted to the fiber of π over each point $P' \in (X'_{U'})^{T'}(\mathbb{C})$ this $\mathcal{O}_E$-action coincides with the CM structure on $A_{P'}$ given by the isogeny $A_{P'} \sim A_{\Phi_1} \times A_{\Phi_2}$.

Local systems in the special case. The existence of the universal abelian variety $A_{U'}$ in the special case allows us to construct the local system $H_1^{\mathrm{dR}}(A_{U'})$ of $E \otimes \mathcal{O}_{X'_{U'}}$-modules on the Shimura curve $X'_{U'}$. This local system is equipped with an integrable connection ∇ and a Hodge filtration

$$0 \to \Omega(A_{U'}^t) \to H_1^{\mathrm{dR}}(A_{U'}) \to \Omega(A_{U'})^\vee \to 0,$$

where $\Omega(A_{U'}) := \pi_*(\Omega_{A_{U'}/X'_{U'}})$ and $\Omega(A_{U'}^t) := \pi_*(\Omega_{A_{U'}^t/X'_{U'}})$. Note that each bundle in this sequence has an E-action induced by the $\mathcal{O}_E$-action on $A_{U'}$.

Since $X'_{U'}$ is defined over $F' \supseteq \tau(F)$, we can construct a morphism $F \otimes \mathcal{O}_{X'_{U'}} \to \mathcal{O}_{X'_{U'}}$ by $f \otimes x \mapsto \tau(f)x$. By abuse of notation we call this morphism τ. Using τ, define the bundles

$$M(A_{U'}, \tau) := H_1^{\mathrm{dR}}(A_{U'}) \otimes_{F \otimes \mathcal{O}_{X'_{U'}}, \tau} \mathcal{O}_{X'_{U'}} \quad \text{and}$$

$$W(A_{U'}, \tau) := \Omega(A_{U'}) \otimes_{F \otimes \mathcal{O}_{X'_{U'}}, \tau} \mathcal{O}_{X'_{U'}}.$$

These sit in the exact sequence

$$0 \to W(A_{U'}^t, \tau^c) \to M(A_{U'}, \tau) \to W(A_{U'}, \tau)^\vee \to 0.$$

We apply the Gauss-Manin connection ∇ to this exact sequence to get the chain of morphisms

$$W(A_{U'}^\vee, \tau^c) \to M(A_{U'}, \tau) \xrightarrow{\nabla} M(A_{U'}, \tau) \otimes \Omega_{X'_{U'}} \to W(A_{U'}, \tau)^\vee \otimes \omega_{X'_{U'}}.$$

Note that we have used the fact that $X'_{U'}$ is a curve (so that $\omega_{X'_{U'}} = \Omega_{X'_{U'}}$). By the Kodaira-Spencer isomorphism, this induces the isomorphism of $E \otimes \mathcal{O}_{X'_{U'}}$-line bundles

$$W(A_{U'}^t, \tau^c) \to W(A_{U'}, \tau)^\vee \otimes \omega_{X'_{U'}}.$$

Taking the determinant gives a further isomorphism of $\mathcal{O}_{X'_{U'}}$-modules

$$N(A_{U'}, \tau) := \det W(A_{U'}, \tau) \otimes \det W(A_{U'}^t, \tau^c) \xrightarrow{\sim} \omega_{X'_{U'}}^{\otimes 2}. \tag{4.36}$$

Every bundle in (4.36) is defined over F'.

To use the line bundle $N(A_{U'}, \tau)$ on $X'_{U'}$ to compute $h([A_0], \tau)$, we need to endow it with a metric and give it an integral structure. While the former has a solution, the latter raises some complications:

1. For each infinite place τ' of F', we can put an explicit metric on $\omega_{X'_{U'}}$ by the formula:

$$\|dz\| = 2\mathrm{Im}(z), \tag{4.37}$$

where z is the coordinate coming from the complex uniformization of $X'_{U',\tau'}$ induced by (4.34). This induces a metric on $N(A_{U'},\tau)_{\tau'}$ for each τ' via the isomorphism (4.36). We point out that this metric on $N(A_{U'},\tau)$ is compatible with one given by Hodge theory. See [60, Thm. 3.7].

2. There is no natural way to extend the moduli problem to $\mathcal{O}_{F'}$. In order to construct an integral model of $X'_{U'}$ over $\mathcal{O}_{F'}$, we need to use the Shimura curve X of quaternion type and Carayol's work [7] on the integral model of X.

4.2.6   Quaternionic Shimura curve X

Constructing the quaternionic Shimura curve.   Now we study the quaternionic Shimura curve X. The motivation for considering this curve was explained at the beginning of §4.2.5. We continue to let E be a CM field such that $[E:\mathbb{Q}]=2g$ and $F \subseteq E$ be its maximal totally real subfield.

Let Σ be a set of places of F of odd cardinality such that it contains all archimedean places of F. Let $\mathbb{B}$ be the quaternion algebra over $\mathbb{A}_F$ whose ramification set is Σ.

For any $\tau \in \operatorname{Hom}(F,\mathbb{C})$, we construct Shimura curves $X_{U,\tau}$ from the following Shimura datum. Let $B=B(\tau)$ be the quaternion algebra over F whose ramification set is $\Sigma \setminus \{\tau\}$. Let $G=\operatorname{Res}_{F/\mathbb{Q}}(B^{\times})$ be a reductive group over $\mathbb{Q}$. Recall that we constructed the following morphism in (4.32)

$$h\colon \mathbb{C}^{\times} \to G(\mathbb{R}) \xrightarrow[\sim]{(\tau,\iota_2,\dots,\iota_g)} \operatorname{GL}_2(\mathbb{R}) \times (\mathbb{H}^{\times})^{g-1},$$

$$z=x+yi \mapsto \left(\begin{pmatrix} x & y \\ -y & x \end{pmatrix}^{-1}, 1, \dots, 1 \right).$$

The $G(\mathbb{R})$-conjugacy class of h is identified with $\mathfrak{h}^{\pm}=\mathbb{C}\setminus\mathbb{R}$ by

$$ghg^{-1} \mapsto g(i), \quad g \in G(\mathbb{R}).$$

The pair $(G,\mathfrak{h}^{\pm})$ satisfies the axioms of a Shimura datum, and so for each $U \subseteq G(\widehat{\mathbb{Q}}) \cong \mathbb{B}_f^{\times}$ we may define a Shimura curve $X_{U,\tau}$ a priori over $\mathbb{C}$ by the uniformization

$$X_{U,\tau}(\mathbb{C}) = G(\mathbb{Q})\backslash \mathfrak{h}^{\pm} \times G(\widehat{\mathbb{Q}})/U. \tag{4.38}$$

In [7], it is shown that each Shimura curve $X_{U,\tau}$ descends to the number field F. We denote by X_U the curve over F thus obtained. The notation above is compatible in the following sense: for each $\tau \in \operatorname{Hom}(F,\mathbb{C})$, the base change of X_U to $\mathbb{C}$ via τ is the one given by (4.38).

Note that when $F \neq \mathbb{Q}$, X_U does not parametrize abelian varieties because the Hodge type induced by h is not $(-1,0)+(0,-1)$. In fact, it is known that X_U is of PEL-type if and only if $F=\mathbb{Q}$ and is compact if and only if $\Sigma \supsetneq \{\tau\}$. While everything holds by working with an appropriate compactification, for technical reasons we will always assume that X_U is compact, or equivalently that Σ is not a singleton.

For the rest of this section we fix an isomorphism $G(\widehat{\mathbb{Q}}) \simeq \mathbb{B}_f^\times$ and a maximal order $\mathcal{O}_\mathbb{B} \subseteq \mathbb{B}$ which contains $\mathcal{O}_{A_E}$. For any N, we define the open subgroup $U(N) := (1 + N\mathcal{O}_\mathbb{B})^\times$ of G.

The integral model. We demonstrate now that there exists a projective system of integral Shimura curves $\mathcal{X}_U$ indexed by open and compact subgroups $U \subseteq \mathcal{O}_\mathbb{B}^\times$. By [60, Prop. 4.1], if $U \subseteq U(N)$ for some $N \geq 3$, we have $g(X_U) \geq 2$, and it is possible to construct a minimal regular (projective flat) model $\mathcal{X}_U$ over the ring of integers $\mathcal{O}_F$.

The Shimura curves X_U form a projective system $\{X_U\}_U$ with the following bonding maps. For each nested pair of open subsets $U' \subseteq U$ of G, there is a finite surjective morphism $X_{U'} \to X_U$ given on the complex points by

$$X_{U',\tau}(\mathbb{C}) = G(\mathbb{Q}) \backslash \mathfrak{h}^\pm \times G(\widehat{\mathbb{Q}})/U' \to G(\mathbb{Q}) \backslash \mathfrak{h}^\pm \times G(\widehat{\mathbb{Q}})/U = X_{U,\tau}(\mathbb{C}).$$

Moreover, if U' is normal in U, then $X_U = X_{U'}/(U/U')$ and the bonding map agrees with the quotient map. To extend this projective system to $\{\mathcal{X}_U\}_U$ is not, however, automatic as without additional hypotheses the projection $X_{U'} \to X_U$ only extends to a rational morphism $\mathcal{X}_{U'} \dashrightarrow \mathcal{X}_U$. To construct a projective system on the integral models, we want this rational map to be an actual morphism.

There is another problem to getting the projective system $\{\mathcal{X}_U\}_U$. As already noted, the minimal regular model $\mathcal{X}_U$ of X_U exists only when $U \subseteq U(N)$ for some $N \geq 3$. Thus, to construct $\mathcal{X}_U$ for arbitrary U, we must first replace U with a smaller open set $U' \subseteq U \cap U(N)$ for some $N \geq 3$ which is normal in U and use the integral model $\mathcal{X}_{U'}$ of $X_{U'}$. The action of U/U' on $X_{U'}$ extends naturally to $\mathcal{X}_{U'}$, and we can define $\mathcal{X}_U$ to be $\mathcal{X}_{U'}/(U/U')$. This seems to solve the problem, but this definition of $\mathcal{X}_U$ does not depend on the choice of U' only if the rational maps in the previous paragraph are actual morphisms.

So we need to put a technical condition on U: we assume $U = \prod_v U_v \subseteq \mathcal{O}_\mathbb{B}^\times$ with $U_v = \mathcal{O}_{\mathbb{B},v}^\times$ for every $v \in \Sigma$. Then Yuan–Zhang [60, Cor. 4.6] showed:

Theorem 4.23. *The system $\{\mathcal{X}_U\}_U$ for such U forms a projective system of curves over $\mathcal{O}_F$ extending the system $\{X_U\}_U$. Moreover, we have*

1. *If $U \subseteq U(N)$ for some $N \geq 3$, then $\mathcal{X}_U$ is smooth at any finite place $v \notin \Sigma$ such that $U_v = \mathcal{O}_{\mathbb{B},v}^\times$ and is a relative Mumford-Tate curve at any finite place $v \in \Sigma$.*

2. *Let H be a finite extension of F which is unramified above any finite place v of F such that $v \in \Sigma$ or $U_v \neq \mathcal{O}_{\mathbb{B},v}^\times$. Then the base change $\mathcal{X}_U \otimes_{\mathcal{O}_F} \mathcal{O}_H$ is $\mathbb{Q}$-factorial, i.e., any Weil divisor has a positive multiple which is Cartier.*

We point out that the proof relies heavily on Carayol's work [7] on the integral model of quaternionic Shimura curves.

We give a sketch for the existence of the projective system. By the discussion preceding this theorem, we may just assume $U \subseteq U(N)$ for some $N \geq 3$.

The most important part of the proof is to understand the local models of X_U. Let p be a prime number and let $\wp$ be a prime of $\mathcal{O}_F$ over p. Fix an ideal $\mathfrak{n} \subseteq \mathcal{O}_F$ which does not contain any prime factor in Σ. Define $U_p(\mathfrak{n}) := (1 + \mathfrak{n}\mathcal{O}_{\mathbb{B},p})^\times$. In this paragraph we localize everything at p or $\wp$, so we may assume that $\mathfrak{n}$ divides a power of p. For any compact open subgroup U^p of $\mathcal{O}_{\mathbb{B}^p}^\times$, use $X_{\mathfrak{n},U^p}$ to denote $X_{U_p(\mathfrak{n}) \times U^p}$. By comparing (the neutral components of) $X_{\mathfrak{n},U^p}$ and some $X'_{\mathfrak{n},U'^p}$ under case (i) of §4.2.5 for some $U'^p \subseteq G'(\widehat{\mathbb{Q}})$, Carayol proved that there is a system of regular models over $\mathcal{O}_\wp$ of $X_{\mathfrak{n},U^p}$ for U^p sufficiently small depending on $\mathfrak{n}$. Moreover this regular model is smooth if $\wp \notin \Sigma$ does not divide $\mathfrak{n}$ and is a relative Mumford curve if $\wp \in \Sigma$ (and hence does not divide $\mathfrak{n}$); see [7, Prop. 4.5.5 and §5.4]. Note that the conclusion here is slightly different from Carayol's original treatment, which only deals with the case where p has no prime factor in Σ (which implies $\wp \notin \Sigma$) and $\wp \nmid \mathfrak{n}$. We refer to Yuan-Zhang [60, Rem. 3.3 and Thm. 4.5] for this difference. The uniqueness of smooth models of curves of genus $g \geq 2$ thus implies that the regular model constructed by Carayol is $X_{\mathfrak{n},U^p} \otimes_{\mathcal{O}_F} \mathcal{O}_\wp$. Now let U^p run over all small U^p so we get a projective system $\{X_{\mathfrak{n},U^p} \otimes_{\mathcal{O}_F} \mathcal{O}_\wp\}_{U^p}$. For an arbitrary $U^p \subseteq (1 + N\mathcal{O}_{\mathbb{B}^p})^\times$ for some $N \geq 3$ prime to p, we can take any U_0^p a sufficiently small normal subgroup of U^p and show that $(X_{\mathfrak{n},U_0^p} \otimes_{\mathcal{O}_F} \mathcal{O}_\wp)/(U^p/U_0^p)$ is isomorphic to $X_{\mathfrak{n},U^p} \otimes_{\mathcal{O}_F} \mathcal{O}_\wp$. See [60, Thm. 4.5] for the computation.

Having the theory of local integral models in hand, it is not hard to give the projective system in Theorem 4.23. For any open compact subgroup $U = \prod_v U_v$ with $U_v = \mathcal{O}_{\mathbb{B},v}^\times$ for any $v \in \Sigma$, we choose a prime number p such that $2 \nmid p$, p has no factors in Σ, and $U_p = \prod_{v|p} \mathcal{O}_{\mathbb{B},v}$. Let $U' = U^p \cdot (1 + p\mathcal{O}_{\mathbb{B},p})^\times$. Then consider the quotient $X_{U'}/(U/U')$. As an important consequence of the theory of local models described in the previous paragraph, the quotient thus defined does not depend on the choice of p. We leave the reader to check that it is the minimal regular model X_U.

Local Kodaira-Spencer map. While X is not of PEL-type in general, its local integral model at a non-archimedean (finite) place parametrizes a family of p-divisible groups. We use this to define a Kodaira-Spencer isomorphism and later use this isomorphism to compute heights.

Let p be a prime number. For any ideal $\mathfrak{n} \subseteq \mathcal{O}_F$ which does not contain any prime factor in Σ, let $U_p(\mathfrak{n}) = (1 + \mathfrak{n}\mathcal{O}_{\mathbb{B},p})^\times$. In particular $U_p(1) = \mathcal{O}_{\mathbb{B},p}^\times$.

Let $X = \varprojlim_U X_U$. Then X has a right action by $G(\widehat{\mathbb{Q}}) = \mathbb{B}_f^\times$ and its maximal subgroup $\overline{F^\times}$, the Zariski closure of $Z(\mathbb{Q}) = F^\times$ in $\mathbb{B}_f^\times$, acts trivially on X. If $F \neq \mathbb{Q}$, then $\overline{F^\times} \neq F^\times$. Define

$$X_\mathfrak{n} := X/U_p(\mathfrak{n}),$$

and the p-divisible group $H_\mathfrak{n}$ on $X_\mathfrak{n}$ by

$$H_\mathfrak{n} := (\mathbb{B}_p/\mathcal{O}_{\mathbb{B},p} \times X)/U_p(\mathfrak{n})$$

where $U_p(\mathfrak{n}) \subseteq \mathcal{O}_{\mathbb{B},p}^\times$ acts on $\mathbb{B}_p/\mathcal{O}_{\mathbb{B},p}$ by right multiplication.

Let $\mathcal{X} = \varprojlim_U \mathcal{X}_U$, where $\{\mathcal{X}_U\}_U$ is the projective system of integral models constructed in Theorem 4.23. Recall that there are some restrictions on U for $\mathcal{X}$, and hence $\mathcal{X}$ is not an integral model of X unless $\mathbb{B}$ splits at all finite places of F. However this will not affect our discussion since $\mathfrak{n}$ is assumed to have no prime factors in Σ.

Now fix a prime $\wp$ of $\mathcal{O}_F$ over p. Let K be the completion of the maximal unramified extension of $F_\wp$ and $\mathcal{O}_K$ its ring of integers.

Let $\mathcal{X}_\mathfrak{n} := \mathcal{X}/U_p(\mathfrak{n})$. Then $\mathcal{X}_\mathfrak{n}$ is an integral model of $X_\mathfrak{n}$. In terms of the local models in the discussion below Theorem 4.23, we have $\mathcal{X}_\mathfrak{n} = \varprojlim_{U^p} \mathcal{X}_{\mathfrak{n},U^p}$. By abuse of notation, we will write $\mathcal{X}_\mathfrak{n}$ for $\mathcal{X}_\mathfrak{n} \otimes_{\mathcal{O}_F} \mathcal{O}_K$ unless otherwise stated. Then $H_\mathfrak{n}$ has an integral model $\mathcal{H}_\mathfrak{n}$ over $\mathcal{O}_K$ with a natural map to $\mathcal{X}_\mathfrak{n}$ extending $H_\mathfrak{n} \to X_\mathfrak{n}$. When $\mathfrak{n} = 1$ we have the further properties: The decomposition $\mathcal{H}_1 = \mathcal{H}_{1,\wp} \times \mathcal{H}_1^\wp$, induced by $\mathcal{O}_{F,p} = \mathcal{O}_\wp \oplus \mathcal{O}_{F,p}^\wp$ as $\mathbb{Z}_p$-algebras, gives the étale-connected decomposition of the p-divisible group. More precisely, $\mathcal{H}_1^\wp$ is étale over $\mathcal{X}_1$, and $\mathcal{H}_{1,\wp}$ is a special formal $\mathcal{O}_{\mathbb{B},\wp}$-module in the sense that $\mathrm{Lie}(\mathcal{H}_{1,\wp})$ is a locally free sheaf over $\mathcal{O}_{\mathcal{X}_{1,\wp}} \otimes_{\mathcal{O}_{F,\wp}} \mathcal{O}_{K_0}$ of rank 1 where K_0 is an unramified quadratic extension of $F_\wp$ embedded into $\mathbb{B}_\wp$; see Yuan-Zhang [60, Thm. 4.9]. We point out that the proof of this extension and decomposition uses Carayol's comparison between $X_\mathfrak{n}$ and some $X_\mathfrak{n}'$ under case (i) of §4.2.5.

For general $\mathfrak{n}$, we write $\mathfrak{n}$ as a product of prime ideals of $\mathcal{O}_F$. By Yuan-Zhang [60, Thm. 4.9.(3)], the morphism $\mathcal{X}_\mathfrak{n} \to \mathcal{X}_1$ classifies certain level structures on $\mathcal{H}_1^\wp$ and $\mathcal{H}_{1,\wp}$, respectively. Hence for the purpose of studying the Kodaira-Spencer map at $\wp$, it suffices to understand the case $\mathfrak{n} = 1$.

Write $\mathcal{W}_{1,\wp} = \mathrm{Lie}(\mathcal{H}_1)^\vee$ and $\mathcal{W}_{1,\wp}^t = \mathrm{Lie}(\mathcal{H}_1^t)^\vee$. The Hodge filtration on the (covariant) Dieudonné crystal of $\mathcal{H}_{1,\wp}$ induces a short exact sequence

$$0 \to \mathcal{W}_{1,\wp}^t \to \mathbb{D}(\mathcal{H}_{1,\wp}) \to \mathcal{W}_{1,\wp}^\vee \to 0.$$

Denote by $\mathcal{X}_{1,\wp} := \mathcal{X}_1 \otimes_{\mathcal{O}_K} \mathcal{O}_{K,\wp}$. Similar to the archimedean case, we apply the Gauss-Manin connection ∇ on $\mathbb{D}(\mathcal{H}_{1,\wp})$ to obtain

$$\mathcal{W}_{1,\wp}^t \to \mathbb{D}(\mathcal{H}_{1,\wp}) \xrightarrow{\nabla} \mathbb{D}(\mathcal{H}_{1,\wp}) \otimes \Omega_{\mathcal{X}_{1,\wp}} \to \mathcal{W}_{1,\wp}^\vee \otimes \omega_{\mathcal{X}_{1,\wp}}.$$

Note that we have used the fact that $\mathcal{X}_{1,\wp}$ is a curve defined over $\mathcal{O}_{K,\wp}$ (so that $\omega_{\mathcal{X}_{1,\wp}} = \Omega_{\mathcal{X}_{1,\wp}}$). This induces a morphism

$$\det \mathcal{W}_{1,\wp}^t \otimes \det \mathcal{W}_{1,\wp} \to \omega_{\mathcal{X}_{1,\wp}}^{\otimes 2}.$$

Denote by $\mathcal{N}_{1,\wp} := \det \mathcal{W}_{1,\wp}^t \otimes \det \mathcal{W}_{1,\wp}$. By the deformation theory of p-divisible groups (or special formal modules) and explicit computation, we can prove that this morphism induces an isomorphism of line bundles on $\mathcal{X}_\wp$ (see [60, Thm. 4.10])

$$\mathrm{KS}_\wp : \mathcal{N}_{1,\wp} \xrightarrow{\sim} \omega_{\mathcal{X}_{1,\wp}}^{\otimes 2}(-\partial_{\mathbb{B},\wp})$$

where

$$\partial_{\mathbb{B},\wp} = \begin{cases} 0 & \text{if } \wp \notin \Sigma, \\ \wp & \text{if } \wp \in \Sigma. \end{cases}$$

Thus, by pulling back under $\mathcal{X}_\mathfrak{n} \to \mathcal{X}_1$, we get

$$\mathrm{KS}_\wp : \mathcal{N}_{\mathfrak{n},\wp} \xrightarrow{\sim} \omega_{\mathcal{X}_{\mathfrak{n},\wp}}^{\otimes 2}(-\partial_{\mathbb{B},\wp}). \tag{4.39}$$

Arithmetic Hodge bundle. For any scheme S, denote by $\mathcal{P}ic(S)$ the groupoid of line bundles on S, and by $\mathcal{P}ic(S)_{\mathbb{Q}}$ the groupoid of objects of the form aL with $a \in \mathbb{Q}$ and $L \in \mathcal{P}ic(S)$. The objects of $\mathcal{P}ic(S)_{\mathbb{Q}}$ are called $\mathbb{Q}$-*line bundles* on S. Similarly we define the groupoid $\widehat{\mathcal{P}ic}(S)_{\mathbb{Q}}$ of hermitian $\mathbb{Q}$-line bundles on an arithmetic variety S. We usually write the tensor products of (hermitian) line bundles additively.

For each open compact subgroup U of $\mathbb{B}_f$, we have a specific line bundle $L_U \in \mathcal{P}ic(X_U)_{\mathbb{Q}}$ called the *Hodge bundle* determined by the following two conditions:

(i) The system $\{L_U\}_U$ is compatible with pullbacks.
(ii) One has $L_U = \omega_{X_U/F}$ if $U \subseteq U(N)$ for some $N \geq 3$.

If we denote by $X = \varprojlim_U X_U$, then we have the following explicit formula for general U:

$$L_U = \omega_{X_U/F} + \sum_{Q \in X_U(\overline{F})} (1 - e_Q^{-1})\mathcal{O}(Q)$$

where e_Q is the ramification index of $X \to X_U$.

The Hodge bundle has been demonstrably very useful to study the geometry of the Shimura curve X and automorphic forms on $\mathbb{B}^\times$; see, for instance, the work of Yuan-S. Zhang-W. Zhang on the Gross-Zagier formula [59]. Here we remark that L_U is the $\mathbb{Q}$-line bundle for holomorphic modular forms of weight two.

For our purpose, we need a system of integral models $\{\mathcal{L}_U\}_U$ of L_U over $\mathcal{X}_U/\mathcal{O}_F$, where $\mathcal{X}_U$ is the integral model of X_U given in Theorem 4.23. Moreover, we need to endow $\mathcal{L}_{U,\tau} = L_{U,\tau}$ with a metric for each archimedean place τ of F to make $\{\mathcal{L}_U\}_U$ into a system of hermitian $\mathbb{Q}$-line bundles. We do this as in [60, Thm. 4.7].

Theorem 4.24. *Assume $U = \prod_v U_v$ with $U_v = \mathcal{O}_{\mathbb{B},v}^\times$ for each $v \in \Sigma$. There is a unique system of hermitian $\mathbb{Q}$-line bundles $\{\overline{\mathcal{L}}_U\}_U$ on the arithmetic surface $\mathcal{X}_U$ extending L_U such that the following conditions hold:*

1. We have $\pi^*_{U',U}\overline{\mathcal{L}}_U = \overline{\mathcal{L}}_{U'}$ for any $U' \subseteq U$, where $\pi_{U',U}: \mathcal{X}_{U'} \to \mathcal{X}_U$ is the natural projection extending $X_{U'} \to X_U$.

2. If $U \subseteq U(N)$ for some $N \geq 3$, then for each finite prime $\wp$ of F such that $U_\wp = \mathcal{O}^\times_{\mathbb{B},\wp}$, there is a canonical isomorphism

$$\mathcal{L}_{U,\wp} = \omega_{\mathcal{X}_U} \otimes \mathcal{O}_\wp / \mathcal{O}_\wp.$$

3. For each archimedean place τ of F, the metric is given by the formula

$$\|dz_\tau\|_\tau = 2\mathrm{Im}(z_\tau),$$

where z_τ is the coordinate coming from the complex uniformization of $X_{U,\tau}$ induced by (4.38).

For any ideal $\mathfrak{n} \subseteq \mathcal{O}_F$ which has no prime factors in Σ, we denote by $U(\mathfrak{n}) = (1 + \mathfrak{n}\mathcal{O}_\mathbb{B})^\times$. We denote by $\mathcal{X}(\mathfrak{n})$ for $\mathcal{X}_{U(\mathfrak{n})}$ and $(L(\mathfrak{n}), \mathcal{L}(\mathfrak{n}), \overline{\mathcal{L}}(\mathfrak{n}))$ for $(L_{U(\mathfrak{n})}, \mathcal{L}_{U(\mathfrak{n})}, \overline{\mathcal{L}}_{U(\mathfrak{n})})$, respectively.

We remark that conditions (1) and (3) of Theorem 4.24 are simply definitions and characterizations. Hence to prove Theorem 4.24, it suffices to construct $\overline{\mathcal{L}}(1)$. To do this we take two different prime numbers p and p', construct the desired hermitian $\mathbb{Q}$-line bundle over $\mathcal{X}(1) \otimes \mathcal{O}_F[1/p]$ and $\mathcal{X}(1) \otimes \mathcal{O}_F[1/p']$, respectively, and then glue. This process also uses the theory of local models of $\mathcal{X}_U$ discussed below Theorem 4.23. We refer to [60, Thm. 4.7] for more details.

We close this section by relating the localization of $\overline{\mathcal{L}}(\mathfrak{n})$ to the local Kodaira-Spencer isomorphism (4.39). Define

$$\mathcal{N}(\mathfrak{n}) := \overline{\mathcal{L}}(\mathfrak{n})^{\otimes 2}(-\partial_\mathbb{B}) \tag{4.40}$$

where $\partial_\mathbb{B}$ is the product of all finite primes in Σ. In the following arguments we always assume that $\mathfrak{n}$ is prime to $\partial_\mathbb{B}$.

Let $\wp \subseteq \mathcal{O}_F$ be a prime ideal and assume that $\wp \nmid \mathfrak{n}$. Denote by K the maximal unramified extension of $F_\wp$ and $\mathcal{O}_K$ its ring of integers.

Recall our convention that $\mathcal{X}_\mathfrak{n} = (\mathcal{X}/U_p(\mathfrak{n})) \otimes_{\mathcal{O}_F} \mathcal{O}_K$ so that there is a natural morphism $\mathcal{X}_\mathfrak{n} \to \mathcal{X}(\mathfrak{n}) \otimes_{\mathcal{O}_F} \mathcal{O}_K$. Localizing at $\wp$, we obtain $pr_{\mathfrak{n},\wp}: \mathcal{X}_{\mathfrak{n},\wp} \to \mathcal{X}(\mathfrak{n}) \otimes_{\mathcal{O}_F} \mathcal{O}_{K,\wp}$. Then (4.41) induces a morphism $\mathcal{N}_{\mathfrak{n},\wp} \to pr^*_{\mathfrak{n},\wp}(\mathcal{N}(\mathfrak{n}) \otimes_{\mathcal{O}_F} \mathcal{O}_{K,\wp})$. We prove that this is an isomorphism as follows. Since $\wp \nmid \mathfrak{n}$, we have $U(\mathfrak{n})_\wp = \mathcal{O}^\times_{\mathbb{B},\wp}$. If $U(\mathfrak{n}) \subseteq U(N)$ for some $N \geq 3$, then condition (2) of Theorem 4.24 holds for $U = U(\mathfrak{n})$, and hence via the local Kodaira-Spencer isomorphism (4.39) we have

$$\mathcal{N}_{\mathfrak{n},\wp} \xrightarrow{\sim} pr^*_{\mathfrak{n},\wp}(\mathcal{N}(\mathfrak{n}) \otimes_{\mathcal{O}_F} \mathcal{O}_{K,\wp}). \tag{4.41}$$

Then (4.41) holds for any $\mathfrak{n}$ prime to $\partial_\mathbb{B}$ such that $\wp \nmid \mathfrak{n}$ by applying this isomorphism to $\mathfrak{n}' = \mathfrak{n}p'$ for a suitable number p' and then applying condition (1) of Theorem 4.24.

4.2.7 Link Between X' and X

The goal of this subsection is to explain the link between X' (under case (ii)) and X. We keep the notation that E is a CM field of degree such that $[E:\mathbb{Q}]=2g$ and $F\subseteq E$ is its maximal totally real subfield. Let (Φ_1,Φ_2) be a pair of nearby CM types of E. Denote by $\tau=\tau_i|_F$, where $\tau_i\in\Phi_i\setminus(\Phi_1\cap\Phi_2)$. Then τ is an archimedean place of F.

Let Σ be a set of places of F of odd cardinality which contains all archimedean places of F. Let $\mathbb{B}$ be the quaternion algebra over $\mathbb{A}_F$ whose ramification set is Σ, and let B be the quaternion algebra over F whose ramification set is $\Sigma\setminus\{\tau\}$. We assume $E\subseteq B$, or equivalently $\mathbb{A}_E\subseteq\mathbb{B}$, so that we are in case (ii) of §4.2.5.

Let F' be the reflex field of $\Phi_1+\Phi_2$ defined by (4.29). Recall that $F'\supseteq\tau(F)$. In §4.2.5, we constructed Shimura curves $X'_{U'}$ of PEL-type which are defined over F'. In §4.2.6, we constructed quaternionic Shimura curves X_U defined over F which in general do not parametrize abelian varieties. The advantages and disadvantages of both types of Shimura curves were listed at the beginning of §4.2.5.

The Shimura curve X''. We will construct another Shimura curve X'' which links X' and X. Recall the following $\mathbb{Q}$-group G'' which we defined in §4.2.5: in terms of $\mathbb{Q}$-points $G''=B^\times\times_{F^\times}E^\times$ is the quotient of $B^\times\times E^\times$ by $F^\times$ under the action $a\circ(b,e)=(ba^{-1},ae)$. The $\mathbb{Q}$-group G' defining X' is a subgroup of G''. See (4.31). We constructed two morphisms $h\colon\mathbb{S}(\mathbb{R})\to B^\times(\mathbb{R})$ and $h_E\colon\mathbb{S}(\mathbb{R})\to E^\times(\mathbb{R})$ in (4.32) and (4.33), respectively.

Let $h''\colon\mathbb{S}\to G''_{\mathbb{R}}$ be the composition of $(h,h_E)\colon\mathbb{S}\to B^\times_{\mathbb{R}}\times E^\times_{\mathbb{R}}$ and the natural projection $B^\times_{\mathbb{R}}\times E^\times_{\mathbb{R}}\to G''_{\mathbb{R}}$. Then h'' factors through $G'_{\mathbb{R}}$ and we denoted in §4.2.5 $h'\colon\mathbb{S}\to G'_{\mathbb{R}}$ the induced morphism. In this subsection, we use both symbols $h''\colon\mathbb{S}\to G''_{\mathbb{R}}$ and $h'\colon\mathbb{S}\to G'_{\mathbb{R}}$ in order to distinguish the targets. Note that $(G'')^{\mathrm{der}}=(G')^{\mathrm{der}}=B^1$, so the $G''(\mathbb{R})$-conjugacy class of h'' is identified with the $G'(\mathbb{R})$-conjugacy class of h', and hence is $\mathfrak{h}^\pm$. The pair $(G'',\mathfrak{h}^\pm)$ defines a Shimura datum, whose reflex field is the same as the reflex field of $(G',\mathfrak{h}^\pm)$ because h' and h'' are essentially the same morphism. So the reflex field of $(G'',\mathfrak{h}^\pm)$ is F'. Hence we have Shimura curves $X''_{U''}$ over F' indexed by open compact subgroups $U''\subseteq G''(\widehat{\mathbb{Q}})$ such that

$$X''_{U'',\tau'}(\mathbb{C})=G''(\mathbb{Q})\backslash\mathfrak{h}^\pm\times G''(\widehat{\mathbb{Q}})/U'',\tag{4.42}$$

for any archimedean place τ' of F'.

Take $X''=\varprojlim_{U''}X''_{U''}$. Then X'' is a scheme over F' and $G''(\widehat{\mathbb{Q}})=\mathbb{B}^\times_f\times_{\mathbb{A}^\times_{F,f}}$ $\mathbb{A}^\times_{E,f}$ acts on X'' on the right. Denote by $\overline{Z''}$ the maximal subgroup which stabilizes X''. Then it is possible to express X'' as a double coset

$$X''_{\tau'}(\mathbb{C})=G''(\mathbb{Q})\backslash\mathfrak{h}^\pm\times G''(\widehat{\mathbb{Q}})/\overline{Z''}.$$

Relating X' and X''. Recall the Shimura curves $X'_{U'}$ defined by the Shimura datum $(G', \mathfrak{h}^{\pm})$ under case (ii) of §4.2.5. Each $X'_{U'}$ is defined over F'.

Take $X' = \varprojlim_{U'} X'_{U'}$. Then X' is a scheme over F' and $G'(\widehat{\mathbb{Q}})$ acts on X' on the right. It can be shown that the maximal subgroup which stabilizes X' is trivial. Hence it is possible to express X' as

$$X'_{\tau'}(\mathbb{C}) = G'(\mathbb{Q})\backslash \mathfrak{h}^{\pm} \times G'(\widehat{\mathbb{Q}}).$$

See Carayol [7, §3.1].

The inclusion $G' \subseteq G''$ induces an F'-morphism

$$i\colon X' \to X''. \tag{4.43}$$

Moreover i is an immersion by the expressions of double cosets of X'' and X' above.

Relating X and X''. Recall that $h''\colon \mathbb{S} \to G''_{\mathbb{R}}$ is the composition of $(h, h_E)\colon \mathbb{S} \to B_{\mathbb{R}}^{\times} \times E_{\mathbb{R}}^{\times}$ and the natural projection $B_{\mathbb{R}}^{\times} \times E_{\mathbb{R}}^{\times} \to G''_{\mathbb{R}}$. At the archimedean place τ of F, the Shimura curve X is given by the Hodge structure defined by h. To relate X and X'', we need to introduce the Shimura variety defined by h_E.

The algebraic group $E^{\times}$ is commutative. Hence the conjugation class of $h_E\colon \mathbb{S} \to E_{\mathbb{R}}^{\times}$ is a singleton which we denote by $\{*\}$. Thus we obtain a Shimura datum $(E^{\times}, \{*\})$. The reflex field of this Shimura datum is F'. Now for any open compact subgroup J of $E^{\times}(\widehat{\mathbb{Q}})$, we have a Shimura variety of dimension zero defined over F'. In fact we have

$$Y_{J,\tau'}(\mathbb{C}) = E^{\times}\backslash \{*\} \times E^{\times}(\widehat{\mathbb{Q}})/J = E^{\times}\backslash E^{\times}(\widehat{\mathbb{Q}})/J \tag{4.44}$$

for any archimedean place τ' of F'.

Take $Y = \varprojlim_{J} Y_J$. Then Y is a scheme over F' and $E^{\times}(\widehat{\mathbb{Q}})$ acts on Y. We have

$$Y_{\tau'}(\mathbb{C}) = \overline{E^{\times}}\backslash E^{\times}(\widehat{\mathbb{Q}}),$$

for any archimedean place τ' of F' where $\overline{E^{\times}}$ is the closure of $E^{\times}$ in $E^{\times}(\widehat{\mathbb{Q}})$.

Now we are ready to relate X and X''. Recall $\tau(F) \subseteq F'$. Let τ' be an archimedean place of F'. Take open compact subgroups U of $\mathbb{B}_f^{\times} = B^{\times}(\widehat{\mathbb{Q}})$ and J of $E^{\times}(\widehat{\mathbb{Q}})$, respectively, and let U'' be the image of $U \times J$ under the projection $(B^{\times} \times E^{\times})(\widehat{\mathbb{Q}}) \to G''(\widehat{\mathbb{Q}})$. Then we obtain a surjective morphism

$$(X_U \times_F Y_J)_{\tau'} \to X''_{U'',\tau'}$$

from the natural projection $B^{\times} \times E^{\times} \to G''$ and the double coset expressions (4.38), (4.44), and (4.42). It can be verified that this morphism descends to F'.

Taking projective limits, we then obtain a morphism of schemes over F'

$$X \times_F Y \to X''.$$

Moreover this morphism is compatible with the right actions of $G(\widehat{\mathbb{Q}}) = \mathbb{B}_f^\times$ on X, $E(\widehat{\mathbb{Q}})$ on Y, and $G''(\widehat{\mathbb{Q}})$ on X''. Thus if we denote by Δ the twisted diagonal map

$$\Delta \colon F^\times \to B^\times \times E^\times, \quad z \mapsto (z, z^{-1}),$$

then we obtain an isomorphism of F'-schemes

$$f \colon (X \times_F Y)/\Delta(F^\times(\widehat{\mathbb{Q}})) \xrightarrow{\sim} X''. \tag{4.45}$$

Since Y has dimension 0, f induces an immersion $X \to X''$ for any point $y \in Y$.

Integral models. Let $U = \mathcal{O}_{\mathbb{B}}^\times$. Take J to be the maximal open compact subgroup $\mathcal{O}_E \otimes_{\mathbb{Z}} \widehat{\mathbb{Z}}$ of $E^\times(\widehat{\mathbb{Q}})$. Then U'', the image of $U \times J$ under the projection $(B^\times \times E^\times)(\widehat{\mathbb{Q}}) \to G''(\widehat{\mathbb{Q}})$, is a maximal open compact subgroup of $G''(\widehat{\mathbb{Q}})$. Then $U' = U'' \cap G'(\widehat{\mathbb{Q}})$ is a maximal open compact subgroup of $G'(\widehat{\mathbb{Q}})$. We shall consider X_U, $X''_{U''}$, and $X'_{U'}$. In terms of §4.2.6, $X_U = X_1$.

We have an integral model $\mathcal{X}_U$ of X_U defined over $\mathcal{O}_F$ by Theorem 4.23. On the other hand, the 0-dimensional variety Y_J has an integral model $\mathcal{Y}_J$ defined over $\mathcal{O}_{F'}$. Hence we obtain an integral model $\mathcal{X}''_{U''}$ of $X_{U''}$ from $\mathcal{X}_U$, $\mathcal{Y}_J$ and (4.45). Thus we obtain an integral model $\mathcal{X}'_{U'}$ of $X_{U'}$ from the immersion (4.43).

Moreover, the Hodge line bundle $\mathcal{L}_U$ induces bundles $\mathcal{L}''_{U''}$ and $\mathcal{L}'_{U'}$ on $\mathcal{X}''_{U''}$ and $\mathcal{X}'_{U'}$, respectively. We use $L''_{U''}$ and $L'_{U'}$ to denote their generic fibers on $X''_{U''}$ and $X'_{U'}$, respectively. These line bundles will be used in the next subsection.

p-divisible groups. Fix a prime number p. We introduce various p-divisible groups. The goal is to compare certain integral structures in the next subsection. Let $U'_p(1)$ be the stabilizer of $\mathcal{O}_{B,p}$ as a left $\mathcal{O}_{B,p}$-module, and let $U''_p(1)$ be the stabilizer of $\mathcal{O}_{\mathbb{B},p}$ as a left $\mathcal{O}_{\mathbb{B},p}$-module.

The first p-divisible group is $H = (\mathbb{B}_p/\mathcal{O}_{\mathbb{B},p} \times X)/\mathcal{O}_{\mathbb{B},p}^\times$ over $X_1 = X/\mathcal{O}_{\mathbb{B},p}^\times$ we introduced in §4.2.6.

The second p-divisible group is $H' := (B_p/\mathcal{O}_{B,p} \times X')/U'_p(1)$ over $X'_1 := X'/U'_p(1)$.

The third p-divisible group is $H'' := (\mathbb{B}_p/\mathcal{O}_{\mathbb{B},p} \times X'')/U''_p(1)$ over $X''_1 := X''/U''_p(1)$.

Let $\wp'$ be a finite place of F' dividing p, and let $\wp$ be a place of F under $\wp'$. Let K', resp. K, be the completion of the maximal unramified extension of $F'_{\wp'}$, resp. $F_\wp$.

We have defined an integral model $\mathcal{H}/\mathcal{X}_{1,\wp}$ in §4.2.6. We can obtain integral models $\mathcal{X}'_{1,\wp'}$ and $\mathcal{X}''_{1,\wp'}$ of X'_1 and X''_1 as before. We would like to extend H' and H'' to integral models $\mathcal{H}'$ and $\mathcal{H}''$. To do this we use the Breuil-Kisin theory. See [60, Prop. 5.2].

4.2.8 Proof of Theorem 4.21

The goal of this subsection is to prove Theorem 4.21.

We start with the right-hand side of Theorem 4.21. We thus retain the following notation. Let E be a CM field of degree $2g$ and let $F \subseteq E$ be its maximal totally real subfield. Let Σ be a set of places of F of odd cardinality which contains all archimedean places of F. Let $\mathbb{B}$ be the quaternion algebra over $\mathbb{A}_F$ whose ramification set is Σ. We assume $\mathbb{A}_E \subseteq \mathbb{B}$. Let X be the Shimura curve defined over F constructed in §4.2.6. It is the projective limit of a projective system of curves over F. Recall that $X = \varprojlim_U X_U$ has a right action by $\mathbb{B}_f^\times$. We consider the action of $E^\times$ on X given by the inclusion $E^\times \subseteq \mathbb{B}_f^\times$.

Let $P \in X^{E^\times}$, $U = \mathcal{O}_\mathbb{B}^\times$, and P_U is the image of P under the projection $X \to X_U$. We have an integral model $\mathcal{X}_U$ of X_U and an arithmetic Hodge line bundle $\overline{\mathcal{L}}_U$ on $\mathcal{X}_U$ given by Theorem 4.23 and Theorem 4.24, respectively.

By (4.26), it suffices to establish an appropriate equality between $h_{\overline{\mathcal{L}}_U}(P_U)$ and $h(\Phi_1, \Phi_2)$ for each nearby CM type (Φ_1, Φ_2) of E. More precisely it suffices to prove

$$g \cdot h(\Phi_1, \Phi_2) = \frac{1}{2} h_{\overline{\mathcal{L}}_U}(P_U) - \frac{1}{4} \log(d_\mathbb{B})$$

for each nearby CM type (Φ_1, Φ_2) of E. Then by definition of $\overline{\mathcal{N}}_U$ (4.40), it suffices to prove

$$4g \cdot h(\Phi_1, \Phi_2) = h_{\overline{\mathcal{N}}_U}(P_U). \tag{4.46}$$

For each nearby CM type (Φ_1, Φ_2) of E, we construct a Shimura curve X' of type $\Phi_1 + \Phi_2$ as in §4.2.5. Note that our assumption $\mathbb{A}_E \subseteq \mathbb{B}$ implies that we are under case (ii) of §4.2.5. We also construct a Shimura curve X'' as in §4.2.7 to link X and X'. We will use the notation introduced in §4.2.5, §4.2.6, and §4.2.7. In particular we use τ to denote the unique archimedean place of F such that $\tau = \tau_i |_F$ where $\tau_i \in \Phi_i \setminus (\Phi_1 \cap \Phi_2)$.

Let Y be as defined under (4.44). Fix a point $y \in Y$, and let P'' be the image of (P, y) under the projection $X \times Y \to X''$. Then P'' is fixed by the subgroup $E^\times \times_{F^\times} E^\times$ of $G''(\mathbb{Q})$.

For a suitable choice of y, P'' is the image of some point $P' \in X'$ under the immersion $X' \to X''$ defined in (4.43). The point P' is then fixed by $T'(\mathbb{Q})$ for the group T' defined in Remark 4.22. See Yuan-Zhang [60, Lem. 5.6].

The maximal open compact subgroup U of $\mathbb{B}_f^\times$ gives rise to a maximal open compact subgroup U'' of $G''(\widehat{\mathbb{Q}})$. Then $U' = U'' \cap G'(\widehat{\mathbb{Q}})$ is a maximal open compact subgroup of $G'(\widehat{\mathbb{Q}})$. Hence the immersions $X \to X''$ (given by f in (4.45) and $y \in Y$) and $X' \to X''$ induce morphisms

$$X_U \to X''_{U'''}, \quad X'_{U'} \to X''_{U'''}. \tag{4.47}$$

Under these morphisms we have $P_U \mapsto P''_{U'''}$, $P'_{U'} \mapsto P''_{U'''}$. The line bundles L_U, $L''_{U'''}$, and $L'_{U'}$ on X_U, $X''_{U'''}$, and $X'_{U'}$, which we defined at the end of §4.2.7, are also compatible under these morphisms.

We have explained in Remark 4.22 that the abelian variety represented by $P'_{U'}$ satisfies the desired properties for the A_0 in (4.27). Hence we can take A_0 to be the abelian variety represented by $P'_{U'}$. Then (4.46) becomes

$$2g \cdot h(A_0, \tau) = h_{\overline{\mathcal{N}}_U}(P_U). \tag{4.48}$$

It suffices now to prove (4.48). Let us understand both sides of it. The abelian variety A_0 is defined over some finite extension K of $F'(P'_{U'})$. Up to enlarging K we may assume $K \supseteq F(P_U)$. The τ-part height $h(A_0, \tau)$ is defined by (4.22). Note that we view A_0 as an abelian variety with an action by $\mathcal{O}_F$, so the field E in (4.22) is F in the current situation. Recall that $\tau(F) \subseteq F'$, hence there is a natural embedding $\tau \colon F \to K$. Thus from (4.22) we get

$$2g \cdot h(A_0, \tau) = \frac{g}{[K:\mathbb{Q}]} \widehat{\deg}(\overline{\mathcal{N}}(A_0, \tau)) \tag{4.49}$$

for the hermitian line bundle $\overline{\mathcal{N}}(A_0, \tau)$ over $\mathrm{Spec}(\mathcal{O}_K)$ defined above (4.22).

The right-hand side of (4.48) is defined as follows. The point P_U defined a morphism $P_U \colon \mathrm{Spec}(F(P_U)) \to X_U$. By Theorem 4.23, X_U has an integral model $\mathcal{X}_U$, and the morphism P_U lift uniquely to a morphism $\overline{P}_U \colon \mathrm{Spec}(\mathcal{O}_{F(P_U)}) \to \mathcal{X}_U$ since $\mathcal{X}_U$ is proper. Then

$$h_{\overline{\mathcal{N}}_U}(P_U) = \frac{1}{[F(P_U):F]} \widehat{\deg}(\overline{P}_U^* \overline{\mathcal{N}}_U). \tag{4.50}$$

Observe that $[F:\mathbb{Q}] = g$. So by (4.49) and (4.50), (4.48) becomes

$$\widehat{\deg}(\overline{\mathcal{N}}(A_0, \tau)) = [K:F(P_U)] \widehat{\deg}(\overline{P}_U^* \overline{\mathcal{N}}_U).$$

Thus it suffices to prove the following isomorphism of hermitian line bundles over $\mathrm{Spec}(\mathcal{O}_K)$:

$$\overline{\mathcal{N}}(A_0, \tau) \cong \overline{P}_U^* \overline{\mathcal{N}}_U \otimes_{\mathcal{O}_{F(P_U)}} \mathcal{O}_K. \tag{4.51}$$

We prove this isomorphism in the following way. First, we look at the generic fibers of both sides. By construction, A_0 is parametrized by $P'_{U'} \in X'_{U'}$. So we can restrict (4.36) to $P'_{U'}$ and get the following conclusion: the generic fiber of $\overline{\mathcal{N}}(A_0, \tau)$ is $(P'_{U'})^* L'^{\otimes 2}_{U'} \otimes_{F'(P'_{U'})} K$.[4] Here we also use $P'_{U'}$ to denote

---

[4]Here we make the following remark: (4.36) holds for U' small enough, whereas the U' in the current situation is the maximal one. Hence in practice we should first apply (4.36) to a suitable subgroup of U' and then use the invariance under pullback. We leave the details for the readers to check.

the morphism $\operatorname{Spec}(F'(P'_{U'})) \to X'_{U'}$. On the other hand, the generic fiber of $\overline{P}^*_U \mathcal{N}_U \otimes_{\mathcal{O}_{F(P_U)}} \mathcal{O}_K$ is by definition $P^*_U L^{\otimes 2}_U \otimes_{F(P_U)} K$.

We explain in (4.47) how P_U and $P'_{U'}$, resp. L_U and $L'_{U'}$, are related via X''. Thus both sides of (4.51) have the same generic fiber $(P''_{U''})^* L''^{\otimes 2}_{U''} \otimes_{F'(P''_{U''})} K$.

Next, we show that both sides of (4.51) have the same metric at the archimedean places. The metric of the left-hand side is given by Hodge theory, and the metric of the right-hand side is defined by part 3 of Theorem 4.24. They coincide by the remark (4.37) at the end of §4.2.5.

Finally, we are left to show that the integral structures defined by both sides of (4.51) on $(P''_{U''})^* L''^{\otimes 2}_{U''} \otimes_{F'(P''_{U''})} K$ are the same. Let v be a finite place of K with residue characteristic p. Let $\mathcal{O}^{\mathrm{ur}}_{K,v}$ be the completion of the maximal unramified extension of $\mathcal{O}_{K,v}$.

The local structure at v of the right-hand side of (4.51) is discussed in §4.2.6. We use the notation in §4.2.6. More precisely, in order to understand the local structure, we only need to work on $\mathcal{X}_{1,v}$ where $\mathcal{X}_1 = \mathcal{X}/U_p(1)$. Then by (4.41) and the definition of $\mathcal{N}_{1,\wp}$ (above (4.39)), the integral structure at v of (4.51) is

$$\overline{P}^*_{U,v}(\det \mathcal{W}^t_{1,v} \otimes \det \mathcal{W}_{1,v}).$$

We note that $\mathcal{W}_{1,v}$ comes from the p-divisible group $\mathcal{H}_1$ on $\mathcal{X}_1$ which we defined in §4.2.6.

The local structure of v of the left-hand side of (4.51) is given by the localization of its definition (4.21). Now to compare the two local structures, we also use X''. First we use the canonical isomorphism

$$\Omega(A_0) \otimes \mathcal{O}^{\mathrm{ur}}_{K,v} \cong \Omega(A_0[p^\infty]) \otimes \mathcal{O}^{\mathrm{ur}}_{K,v}$$

to pass from A_0 to its p-divisible group. Hence we can use the p-divisible group $\mathcal{H}'/\mathcal{X}'_{1,v}$, which is introduced at the end of §4.2.7, to understand the local structure of $\overline{\mathcal{N}}(A_0, \tau)$ at v.

Then the local structures at v of both sides of (4.51) already looks alike, except that we now need to compare the p-divisible groups $\mathcal{H}_1/\mathcal{X}_{1,v}$ and $\mathcal{H}'_1/\mathcal{X}'_{1,v}$. This comparison is done via the p-divisible group $\mathcal{H}''_1/\mathcal{X}''_{1,v}$. We need the Breuil-Kisin theory in the comparison. We omit the computation and refer to [60, Prop. 5.1, Prop. 5.3, Prop. 5.4, Cor. 5.5] for details.

4.3 THE GENERALIZED CHOWLA-SELBERG FORMULA

In this section we discuss the generalized Chowla-Selberg formula in Yuan-Zhang [60]. This formula, together with the main result of the previous section (Theorem 4.21), directly implies the averaged Colmez conjecture in Theorem 4.14. After stating the result, we explain various aspects of the strategy used in the proof; see §4.3.1 for an outline.

To state the formula, let E be an arbitrary CM field with maximal totally real subfield $F \subseteq E$ and let $\mathbb{B}$ be a totally definite incoherent quaternion algebra over $\mathbb{A}_F$. Suppose that there exists an embedding $\mathbb{A}_E \hookrightarrow \mathbb{B}$ over $\mathbb{A}_F$. Let X be the quaternionic Shimura curve constructed in §4.2.6 and consider a point $P \in X$ which is fixed by $E^\times$. We assume that $U = \prod U_v$ is a maximal open compact subgroup of $\mathbb{B}_f^\times$ such that U contains $\widehat{\mathcal{O}}_E^\times$. Then we denote by $h_{\overline{\mathcal{L}}_U}(P_U)$ the height of $P_U \in X_U$ defined in (4.13), where P_U is the image of P in $X_U = X/U$. We now can state the generalized Chowla-Selberg formula obtained in [60, Thm. 1.7].

Theorem 4.25. *Assume that at least two places of F are ramified in $\mathbb{B}$, and suppose that there is no finite place of F which ramifies both in E and $\mathbb{B}$. Then it holds*

$$h_{\overline{\mathcal{L}}_U}(P_U) = -\frac{L'(\eta, 0)}{L(\eta, 0)} + \tfrac{1}{2} \log(\tfrac{\Delta_{\mathbb{B}}}{\Delta_{E/F}}).$$

Here $L'(\eta, s)$ is the derivative of the L-function $L(\eta, s)$ of the quadratic character η of $\mathbb{A}_F^\times / F^\times$ defined by E/F. Further, $\Delta_{E/F}$ is the norm of the relative discriminant of E/F, and $\Delta_{\mathbb{B}}$ is the norm of the product of the prime ideals of F over which $\mathbb{B}$ ramifies.

In the simplest case when the CM field E is quadratic, the right-hand side of the above formula coincides (up to the term $\tfrac{1}{2} \log \Delta_{\mathbb{B}}$) with the right-hand side of the reformulation in Theorem 4.11 of the classical Lerch-Chowla-Selberg formula in terms of the Faltings height.

4.3.1   Strategy of Proof of Theorem 4.25

In the remainder of §4.3 we explain some aspects of the proof of Theorem 4.25 which uses an extension of the method of Yuan-Zhang-Zhang [59]. In fact, one goal of this section is to illustrate that the generalized Chowla-Selberg formula in Theorem 4.25 is a special case of a generalized Gross-Zagier formula coming from a fundamental identity between a geometric and analytic kernel. In the following discussion we give inter alia an outline of §4.3.

Classical Lerch-Chowla-Selberg formula. In §4.3.2 we shall discuss in detail the case of quadratic CM fields. In this case, Theorem 4.25 is essentially the classical Lerch-Chowla-Selberg formula. A result of Faltings shows that this classical formula is equivalent to the Colmez conjecture for quadratic CM fields. After giving a proof of Faltings's result, we explain how one can deduce the classical Lerch-Chowla-Selberg formula from Kronecker's limit formula. We also mention obstructions for generalizing the (classical) methods described in §4.3.2.

Gross-Zagier formula. To deal with general CM fields, we take a different point of view. Instead of computing the height of a CM point directly as done in §4.3.2 for quadratic CM fields, we relate the height to some generating function Z of intersection numbers of special cycles and invoke modularity results. In §4.3.3

we illustrate this strategy by discussing the classical Gross-Zagier formula. In the proof of their formula, Gross-Zagier establish an identity

$$Z = \Phi \tag{4.52}$$

between two modular forms Z and Φ called geometric and analytic kernel, respectively. We shall explain some aspects of the computation of the Fourier coefficients of Z which are given by the Néron-Tate height pairing of two (Heegner) points. In this computation, Gross-Zagier implicitly determine the self-intersection of a CM point which gives via the adjunction formula a new proof of the classical Lerch-Chowla-Selberg formula. In fact, the fundamental identity (4.52) contains the classical Lerch-Chowla-Selberg formula as a special case.

Proof of Theorem 4.25 via Yuan-Zhang-Zhang. In §4.3.4 we will first briefly recall Yuan-Zhang-Zhang's proof of the complete Gross-Zagier formula [59]. It follows the principle that one should compare the geometric kernel Z and the analytic kernel Φ. They constructed Z and Φ, and showed that both are cusp forms of weight 2. Then one studies the difference

$$\Phi - Z.$$

The geometric kernel Z can be split into two parts $Z = Z_{\text{prop}} + Z_{\text{self}}$, according to proper intersection and self-intersection. While proper intersections can be computed, it is not possible to compute Z_{self} directly. Yuan-Zhang-Zhang showed that for the purpose of proving Gross-Zagier, one can choose some nice Schwartz functions such that the essential part of Z_{self} vanishes. Then one can observe that $\Phi - Z$ is a sum of theta series of weight 1. But $\Phi - Z$ is a cusp form of weight 2, since Φ and Z are both cusp forms of weight 2. Therefore it follows that

$$\Phi - Z = 0.$$

Since Z essentially consists of proper intersections which are computable, and since the analytic kernel Φ is computable, one can conclude the complete Gross-Zagier formula.

We finally describe the proof of Theorem 4.25. As mentioned, it is possible to choose some nice Schwartz functions such that the essential part of Z_{self} vanishes. To prove the generalized Chowla-Selberg formula, we need to choose different Schwartz functions since the height $h_{\overline{\mathcal{L}_U}}(P_U)$ in Theorem 4.25 comes from the essential part of Z_{self} by some arithmetic adjunction formula. However, as in the above described proof of the complete Gross-Zagier formula, we still would like to prove that the difference $\Phi - Z$ is a sum of theta series of weight 1. Suppose this is the case. Then $\Phi - Z$ has simultaneously weight 1 and 2, which means that $\Phi - Z = 0$ and thus

$$Z_{\text{self}} = \Phi - Z_{\text{prop}}.$$

This identity allows then to conclude Theorem 4.25, since the functions Z_{self}, Φ, and Z_{prop} can be explicitly computed. Finally, to prove that $\Phi - Z$ is a sum of theta series of weight 1, one uses an extension of the method of [59]. In the setup of the extension, the function $\Phi - Z$ is a sum of series which are a priori *not* automorphic, although they look like theta series. One major new idea in [60] is to study these series, called *pseudo-theta series*. If a sum of pseudo-theta series is automorphic, then one can prove that each series can be replaced by the difference of two associated theta series which are explicit. Then one observes that each theta series obtained in this way is of weight 1, and hence $\Phi - Z$ is a sum of theta series of weight 1 as desired.

4.3.2   Classical Lerch-Chowla-Selberg Formula

In this section we discuss various aspects of the classical Lerch-Chowla-Selberg formula and its formulation in terms of the Faltings height of a CM elliptic curve. Throughout this section, we let $E \subseteq \mathbb{C}$ be an imaginary quadratic number field of discriminant $-d < 0$ and we denote by $\eta : (\mathbb{Z}/d\mathbb{Z})^\times \to \{\pm 1\}$ the quadratic character of the quadratic field extension $E/\mathbb{Q}$.

Colmez conjecture in dimension one.   For any CM type Φ of E, we denote by $h(\Phi)$ its Faltings height defined in (4.6). The following result is (equivalent to) Conjecture 4.13 of Colmez restricted to the case of imaginary quadratic number fields; see also Theorem 4.11.

Theorem 4.26 (Conjecture 4.13 for Quadratic E).   *If Φ is a CM type of E, then*

$$h(\Phi) = -\tfrac{1}{2} \frac{L'(\eta, 0)}{L(\eta, 0)} - \tfrac{1}{4} \log d. \tag{4.53}$$

Here we recall that $L'(\eta, s)$ is the derivative of the Dirichlet L-function $L(\eta, s)$ of η. There exist exactly two distinct CM types of a given imaginary quadratic field and hence (4.53) is the averaged Colmez Conjecture 4.14 for imaginary quadratic number fields. We shall see below that Theorem 4.26 is a reformulation of the classical Lerch-Chowla-Selberg formula in terms of $h(\Phi)$. In fact Theorem 4.26 was already proven before Colmez made his conjecture. For example, it can be found in Deligne's review [14, p. 29] of Faltings's work [19] in which Deligne stated (4.53) with $L'(\eta, 0)/L(\eta, 0)$ expressed in terms of the Γ-function.

Faltings height of elliptic curves.   To see that the Colmez conjecture for imaginary quadratic fields is equivalent to the classical Lerch-Chowla-Selberg formula, we shall use the following result of Faltings [21, Thm. 7.b] which we shall apply with an elliptic curve having CM by $\mathcal{O}_E$.

Theorem 4.27. *Suppose that A is an elliptic curve defined over an arbitrary number field K. Then the Faltings height $h(A)$ of A satisfies*

$$12[K:\mathbb{Q}]h(A) = \log \Delta(A) - \sum \log\big(\Delta(\tau_\sigma)(4\pi\mathrm{im}(\tau_\sigma))^6\big) \qquad (4.54)$$

with the sum taken over all embeddings $\sigma: K \hookrightarrow \mathbb{C}$, where for each embedding $\sigma: K \hookrightarrow \mathbb{C}$ the complex number τ_σ with $\mathrm{im}(\tau_\sigma) > 0$ is chosen such that $(A \times_{K,\sigma} \mathbb{C})(\mathbb{C}) \simeq \mathbb{C}/(\mathbb{Z} + \tau_\sigma \mathbb{Z})$.

Here it is important to recall that Faltings normalizes the metric (4.1) via the factor $\frac{1}{2}$, while we use the factor $\frac{1}{2\pi}$ as in Deligne [14] and Yuan-Zhang [60]. Further, $\Delta(\tau_\sigma)$ is the usual elliptic discriminant modular form given by $\Delta(\tau_\sigma) = q \prod(1-q^n)^{24}$ for $q = \exp(2\pi i\tau_\sigma)$, while $\Delta(A)$ is the stable discriminant of the curve A. The stable discriminant $\Delta(A)$ can be defined by

$$\log \Delta(A) = \tfrac{1}{[L:K]} \sum n_v \log N_v$$

with the sum taken over all finite places v of any number field L containing K such that the minimal regular model $\mathcal{X}$ of $A \times_K L$ over $\mathcal{O}_L$ is semistable, where n_v is the number of singular points of the geometric fiber of $\mathcal{X}$ over v and N_v is the cardinality of the residue field of v. Then $[K:\mathbb{Q}]^{-1} \log \Delta(A)$ and $[K:\mathbb{Q}]^{-1}$ times the right-hand side of (4.54) are stable under base change to an arbitrary finite field extension of K. We used this base change property to reduce Theorem 4.27 to the semistable case which is treated in [21, Thm. 7.b].

Remark 4.28 (Faltings-Noether formula). Faltings's computation of his delta invariant [21, p. 417] and the vanishing of the self-intersection in [21, Thm. 7.a] show that Theorem 4.27 is in fact the so-called Noether formula of Faltings [21, Thm. 6] restricted to the special case of elliptic curves. Moret-Bailly [41, Thm. 2.2] computed the absolute constant in [21, Thm. 6].

Classical Lerch-Chowla-Selberg formula. We first introduce some notation. For any fractional ideal I of E, we write I^{-1} for its inverse fractional ideal and we denote by $\Delta(I)$ its modular discriminant which is defined, for example, in [54, p. 34]. Next we put

$$F(I) = \Delta(I^{-1})\Delta(I).$$

Transformation properties of $\Delta(I)$ assure that $F(I)$ depends only on the class $[I]$ generated by I in the class group $\mathrm{Pic}(\mathcal{O}_E)$ of E. We denote by $h = |\mathrm{Pic}(\mathcal{O}_E)|$ the class number of E and we write ζ_E' for the derivative of the Dedekind zeta function ζ_E of the number field E. Now we can state the classical Lerch-Chowla-Selberg formula, formulated[5] as in [54, p. 91].

---

[5]Here we used that $\zeta_E'(0) = -\frac{h}{w}$ where w denotes the number of roots of unity in E.

Theorem 4.29. *It holds that*

$$\frac{1}{12h} \sum_{[I] \in \mathrm{Pic}(\mathcal{O}_E)} \log F(I) = \frac{\zeta'_E(0)}{\zeta_E(0)}. \tag{4.55}$$

This theorem was essentially proven by Lerch [36] in 1897, and it was then rediscovered by Chowla-Selberg [11, 12]. We remark that the right-hand side of (4.55) can be expressed in terms of the L-function $L = L(\eta, s)$, since it holds that $\zeta_E = \zeta L$ for ζ the classical Riemann zeta function. More precisely, on using that $\zeta'(0) = -\frac{1}{2} \log(2\pi)$ and $\zeta(0) = -\frac{1}{2}$, we obtain

$$\frac{\zeta'_E(0)}{\zeta_E(0)} = \frac{L'(0)}{L(0)} + \log(2\pi). \tag{4.56}$$

Next, we explain the ideas of a proof of Theorem 4.29 which uses Kronecker's limit formula. We suppose that $I = \mathbb{Z} + \tau \mathbb{Z}$ with $\mathrm{im}(\tau) > 0$ and then we consider the Eisenstein series

$$E(\tau, s) = \sum \frac{\mathrm{im}(\tau)^s}{|m\tau + n|^{2s}}$$

with the sum taken over all pairs $(m, n) \in \mathbb{Z}^2$ with $(m, n) \neq 0$. The Eisenstein series $E(\tau, s)$ has a pole at $s = 1$. In fact, the Kronecker limit formula moreover gives

$$E(\tau, s) - \frac{\pi}{s-1} = -\frac{\pi}{6} \log |\Delta(\tau) \mathrm{im}(\tau)^6| + c + O(s-1),$$

where c is an explicit absolute constant. Further, one can relate $E(\tau, s)$ to $\zeta_E(s)$ and the formula (4.58) below expresses $F(I)$ in terms of $|\Delta(\tau) \mathrm{im}(\tau)^6|^2$. This then allows one to deduce Theorem 4.29 from the above displayed Kronecker limit formula. A detailed proof of the classical Lerch-Chowla-Selberg formula via Kronecker's limit theorem can be found in [54].

Proof of Theorem 4.26. As already mentioned, the Colmez conjecture for imaginary quadratic fields is a direct consequence of the classical Lerch-Chowla-Selberg formula (4.55) and Faltings's formula (4.54). To see this, let Φ be a CM type of E and let $A = A_\Phi$ be the elliptic curve in (4.5) which has CM by $(\mathcal{O}_E, \Phi)$. Then A is defined over the number field $\mathbb{Q}(j_A)$ generated by its j-invariant j_A and therefore an application of Theorem 4.27 with A gives

$$h(\Phi) = -\frac{1}{12h} \sum_{\sigma: \mathbb{Q}(j_A) \hookrightarrow \mathbb{C}} \log(\Delta(\tau_\sigma)(4\pi \mathrm{im}(\tau_\sigma))^6). \tag{4.57}$$

Here we used class field theory which provides that $h = [H : E] = [\mathbb{Q}(j_A) : \mathbb{Q}]$ for H the Hilbert class field of E, and we exploited that our A with CM has potentially good reduction everywhere which means that $\Delta(A) = 1$. Next we consider the following identity

$$F(I) = (2\pi)^{12} d^{-6} |\Delta(\tau)|^2 (4\pi \mathrm{im}(\tau))^{12}, \tag{4.58}$$

which holds for any fractional ideal I of E of the form $I = \mathbb{Z} + \tau\mathbb{Z}$ with $\text{im}(\tau) > 0$. Finally, on combining (4.56), (4.57), and (4.58), we see that the Lerch-Chowla-Selberg formula (4.55) is equivalent to the formula (4.53). This completes the proof of Theorem 4.26.

Obstructions to generalizing the method. The above computation of $h(\Phi)$ relies on the explicit construction of a certain modular form on $\overline{\mathcal{M}}$. This obstructs generalizing the formula to other Shimura curves, where something similar to the discriminant is lacking. In particular, without introducing substantial new ideas the above described method does not allow one to deal with more general Shimura curves or with the moduli stacks $\mathcal{A}_g$ for $g > 1$.

Remark 4.30. One can compute the Faltings height of Jacobians of hyperelliptic curves by using modular forms; see S. Zhang's work [63] on the Gross-Schoen cycle and dualizing sheaves.

4.3.3   Gross-Zagier Formula

We next would like to explain the close relation between the classical Lerch-Chowla-Selberg formula and the original Gross-Zagier formula. After stating the original Gross-Zagier formula, we discuss some aspects of its proof. We denote by $\overline{\mathbb{Q}}$ the algebraic closure of $\mathbb{Q}$ in $\mathbb{C}$.

Modular forms and modular curves. To state the formula of Gross-Zagier, we need to introduce some terminology from the theory of modular forms and modular curves. We refer to [16] for an elementary presentation of the theory which is sufficient for our purpose in this section. Let $N \geq 1$ be a rational integer. We denote by ∞ and 0 the usual two distinguished cusps of the smooth projective and geometrically connected modular curve $X_0(N)$ over $\mathbb{Q}$. To simplify notation we write $X = X_0(N)$. The cusps ∞ and 0 are both $\mathbb{Q}$-rational points of X, and the non-cuspidal points of $X(\overline{\mathbb{Q}})$ correspond to diagrams

$$\phi : A \to A' \tag{4.59}$$

where A and A' are elliptic curves over $\overline{\mathbb{Q}}$ and ϕ is an isogeny with kernel isomorphic to $\mathbb{Z}/N\mathbb{Z}$. Let $J = \text{Pic}^0(X)$ be the Jacobian of X. For each integer $m \geq 1$, the m-th Hecke operator T_m acts on J. Further we consider the Hecke algebra $\mathbb{T} = \mathbb{Z}[T_m, m \geq 1]$.

The Gross-Zagier formula. Suppose that E is an imaginary quadratic number field of discriminant $-d < 0$ such that d is coprime to $2N$. We say that a point in $X(\overline{\mathbb{Q}})$ is a Heegner point if it corresponds via (4.59) to a diagram $\phi : A \to A'$ such that the elliptic curves A and A' both have CM by $\mathcal{O}_E$. There exists a Heegner point of $X = X_0(N)$ if and only if d is congruent to a square modulo $4N$. Further, any Heegner point of X lies in $X(H)$ where H is the Hilbert class field of E. Let $S_2(\Gamma_0(N))$ be the space of weight two cusp forms for the modular group $\Gamma_0(N)$, and let $f \in S_2(\Gamma_0(2))$ be a newform. For any point $Q \in J(H) \otimes \mathbb{C}$, let $Q(f)$ be the projection of Q to the f-isotypical component of $J(H) \otimes \mathbb{C}$ under the action of the Hecke algebra $\mathbb{T}$. The following result was established by Gross-Zagier in [28, Thm. 6.1].

Theorem 4.31. *Let $L'(f, s)$ be the derivative of the L-function $L(f, s)$ of f. Suppose that x is a Heegner point of X and consider $P = [x - \infty] \in J(H)$. Then $Q = \sum_{\sigma \in \mathrm{Gal}(H/E)} P^\sigma$ satisfies*

$$L'(f, 1) = * \cdot \langle Q(f), Q(f) \rangle.$$

Here $\langle \cdot, \cdot \rangle$ denotes the Néron-Tate height pairing on $J(H) \times J(H)$, and $*$ is a nonzero complex number which was explicitly computed in [28, Thm. 6.1]. The above theorem of Gross-Zagier has many interesting arithmetic applications; see, for example, [28, §1.7 and §1.8].

Strategy of proof. We now explain the main ideas of the proof of Theorem 4.31. Gross-Zagier first prove an identity between two modular forms, which we call the geometric kernel and the analytic kernel. Then they deduce Theorem 4.31 by applying to this identity the Petersson inner product with our given newform $f \in S_2(\Gamma_0(N))$. To define the geometric kernel, we take arbitrary points P, Q in $J(H)$ and we consider the generating function

$$Z(P, Q) = \sum_{m \geq 1} \langle P, T_m Q \rangle \, q^m.$$

A short formal argument (see [28, p. 306]) shows that the function $Z(P, Q)$ lies in $S_2(\Gamma_0(N))$. Suppose now that x is a Heegner point of X and consider $P = [x - \infty] \in J(H) \otimes \mathbb{C}$. For any given $\sigma \in \mathrm{Gal}(H/E)$, we then define the geometric kernel

$$Z_\sigma = Z(P, P^\sigma).$$

On using holomorphic projection and the Ranking-Selberg method, Gross-Zagier construct a form $\Phi_\sigma \in S_2(\Gamma_0(N))$ whose Petersson inner product with any newform of $S_2(\Gamma_0(N))$ is essentially given by the derivative of a certain L-function evaluated at 1. Furthermore, they show that, after scaling and possibly adding an oldform to Φ_σ, one obtains the identity

$$Z_\sigma = \Phi_\sigma. \tag{4.60}$$

They prove this identity between the geometric kernel Z_σ and the analytic kernel Φ_σ by computing for each form its Fourier coefficients. To discuss an interesting aspect of this computation, we consider $Q = [x - 0] \in J(H)$. For any rational integer $m \geq 1$, we recall that the m-th Fourier coefficient of Z_σ is given by $\langle P, T_m P^\sigma \rangle$

which equals $\langle P, T_m Q^\sigma \rangle$ by the Manin-Drinfeld theorem. The Néron-Tate height pairing $\langle P, T_m Q^\sigma \rangle$ can be studied (see [28, §3]) via the (local) intersection $\bar{x} \cdot \bar{y}$ of two divisors $\bar{x}$ and $\bar{y}$, where $\bar{x}$ denotes the Zariski closure of $x \in X$ in the canonical Deligne-Rapoport integral model $\mathcal{X}$ of X over $\mathbb{Z}$. In the case when $\bar{x} \neq \bar{y}$, one can compute $\bar{x} \cdot \bar{y}$ by using the moduli interpretation of $\mathcal{X}/\mathbb{Z}$. On the other hand, Gross-Zagier formulated a new moduli interpretation to deal with the case when $\bar{x} = \bar{y}$. They obtained a formula for the self-intersection $\bar{x}^2$ in terms of $L'(\eta, 0)/L(\eta, 0)$, which implies Theorem 4.11 by the adjunction formula relating $-\bar{x}^2$ to the Faltings height $h(A)$ of A corresponding to x via (4.59). In other words, Gross-Zagier have implicitly obtained a new proof of the classical Lerch-Chowla-Selberg formula via a new moduli interpretation of the Faltings height.

4.3.4 Proof of Theorem 4.25

Yuan and Zhang's idea to prove Theorem 4.25 in [60] is different from the previous proof of Chowla-Selberg formula. They follow the strategy in the proof of the complete Gross-Zagier formula of Yuan-Zhang-Zhang [59]. The basic principle is that the Chowla-Selberg formula should be a "special case" of the following grand formula of weight 2 cusp forms

$$Z = \Phi. \tag{4.61}$$

Here Z is defined by generating series (and hence in a geometric way) and Φ is the holomorphic projection of the linear combination of products of some modular forms. We call Z the *geometric kernel* and call Φ the *analytic kernel*. Then the geometric kernel Z gives the Néron-Tate height, and the analytic kernel Φ gives the L-function. The identity (4.60) in the proof of the original Gross-Zagier formula is a particular case of (4.61).

Proof of the fundamental identity. Let us give a brief explanation on how to prove the equality (4.61). The definition of Z involves the Néron-Tate pairing and Hecke correspondence. Hence the terms of Z can be regrouped into two parts $Z = Z_{\text{prop}} + Z_{\text{self}}$ according to proper intersection and self-intersection. Proper intersections can be computed but not self-intersection. Although the argument is complicated, the upshot is that one does not have to compute the self-intersection. This uses representation theory and a multiplicity one argument. In fact, as a by-product of this approach, one obtains a formula for the self-intersection $-x^2$ and, thus, a formula for $h_{\overline{\mathcal{L}}(1)}(x)$ via the adjunction formula. In more detail, we can show the following:

(1) Z_{prop} can be explicitly computed so that $\Phi_1 = \Phi - Z_{\text{prop}}$ is a weight 1 form involving the constant coefficients of the Eisenstein series.
(2) Z_{self} is a sum of forms of weight 1 though *cannot be explicitly computed*.

Now we have the fundamental identity

$$\Phi - Z = \Phi_1 - Z_{\text{self}}. \tag{4.62}$$

This is an identity of forms *weight 2 form = weight 1 form* which implies that both sides in (4.62) are zero. In particular, this means that $\Phi = Z$ giving the complete Gross-Zagier formula.

Deduction of the generalized Chowla-Selberg formula. On the other hand, the vanishing of the right-hand side of (4.62) implies that $\Phi_1 = Z_{\text{self}}$. Then we obtain Theorem 4.25 by relating Z_{self} to $h_{\overline{\mathcal{L}}(1)}(x)$ via the arithmetic adjunction formula ([60, Thm. 9.3]) and by computing Φ_1. Here the computation of Φ_1 can be found in the proof of [60, Thm. 9.1].

Finally, we point out that proving that Φ_1 and Z_{self} are sums of weight 1 forms can be a highly nontrivial task. A priori when we do the difference $\Phi - Z$, we get a sum of series which look like theta series but which are *not* automorphic.[6] One major new idea of [60, Part II] is to study these terms. Yuan and Zhang call them *pseudo-theta series*. They proved [60, Lem. 6.1] that if a sum of pseudo-theta series is automorphic, then we can replace each one of them by the difference of two associated theta series which are explicit. It is then not hard to see that the associated theta series are of weight 1 in this case.

The following is an example for part (2) in the modular curve $X_0(1)$. Let $x = [\mathbb{C}/\mathcal{O}_E]$. The multiplicity of x in $T_m(x)$ is

$$\text{mult}_x(T_m(x)) = \#\{\phi \in \text{End}(\mathbb{C}/\mathcal{O}_E), \deg \phi = m\} = \#\{\phi \in \mathcal{O}_E, \text{N}(\phi) = m\} = r_m.$$

Then

$$\sum \langle x, T_m x \rangle_{NT} q^m = \langle x, x \rangle_{NT} \theta_E(q) + \text{terms without self-intersection,}$$

where $\theta_E(q) = \sum r_m q^m$. Note that $\theta_E(q)$ is a weight one modular form.

4.4 HIGHER CHOWLA-SELBERG/GROSS-ZAGIER FORMULA

Throughout this section, fix k to be a finite field of characteristic $p > 2$ and let X be a geometrically connected, smooth, proper curve over k. Let $\nu : X' \to X$ be a connected étale double cover such that X' is also geometrically connected. Let $F = k(X)$ and $F' = k(X')$ denote their function fields, and g and g' be the genera of X and X', respectively. In particular, $g' = 2g - 1$.

For each closed point $x \in |X|$, let $\mathcal{O}_x$ be the completed local ring of X at x and F_x be its fraction field. Define the ring of adeles $\mathbb{A} = \prod'_{x \in |X|} F_x$ which has

---

[6]For the purpose of Gross-Zagier, it is possible to choose some nice Schwartz functions so that every term in the sum is a theta series. See [59, Thm. 5.7]. But then the part of Z_{self} corresponding to $h_{\overline{\mathcal{L}}(1)}(x)$ vanishes. See [59, the end of §5.2.1]. So we cannot make this choice if we want to compute $h_{\overline{\mathcal{L}}(1)}(x)$.

ring of integers $\mathbb{O} = \prod_{x \in |X|} \mathcal{O}_x$. Denote by

$$\eta = \eta_{F'/F} : F^\times \backslash \mathbb{A}^\times / \mathbb{O} \to \{\pm 1\}$$

the character corresponding to the étale double cover X'/X by class field theory.

Our goal is to give a function field analogue of both the Gross-Zagier and Chowla-Selberg formulae for special values of L-functions in the function field setting. The rough version of the theorem is the following (see Theorem 4.47 and Theorem 4.48 in §4.4.3 for precise statements).

Theorem 4.32. *Let π be an everywhere unramified cuspidal automorphic representation of $G = \mathrm{PGL}_2$ over F and let $\theta_*^r[\mathrm{Sht}_T^r]$ be a Heegner-Drinfeld cycle on $\mathrm{Sht}_G^{\prime r}$. Then up to normalization and constant parameters depending on the curve X, we have for even $r \geq 0$ the special value formula*

$$L^{(r)}(\pi_{F'}, 1/2) \doteq \langle \theta_*^r[\mathrm{Sht}_T^r]_\pi, \theta_*^r[\mathrm{Sht}_T^r]_\pi \rangle_{\mathrm{Sht}_G^{\prime r}, \pi}$$

where $\langle , \rangle_{\mathrm{Sht}_G^{\prime r}}$ expresses an intersection pairing on the stack $\mathrm{Sht}_G^{\prime r}$.

The simplest instance of this theorem is that it derives the function field analogue of the *Kronecker limit formula*, which will relate to the special values of the L-function associated to the quadratic character η defined above. First, note that η gives a representation on the module $V = \mathrm{H}^1_{\text{ét}}(X', \mathbb{Q}_\ell) / \mathrm{H}^1_{\text{ét}}(X, \mathbb{Q}_\ell)$, and so its L-function is given by the expansion:

$$L(\eta, s) = \sum_{\mathfrak{a} \in \mathcal{O}_F} \frac{\eta(\mathfrak{a})}{N(\mathfrak{a})^s} = \frac{\zeta_{X'}(s)}{\zeta_X(s)} = \det(\mathrm{Frob}_q \, q^{-s} | V).$$

A priori the final expression indicates that $L(\eta, s)$ is a polynomial in q^{-s} with degree $\dim(V) = 2(g' - g) = 2g - 2$. While not immediately relevant, we do point out that this phenomenon is markedly different from what is expected for number fields. Then, in this instance, Theorem 4.32 is the following.

Theorem 4.33. *When $r \geq 0$ is even, we have*

$$\frac{2^{r+2}}{(\log q)^r} L^{(r)}(\eta, 0) = \langle \theta_*^r[\mathrm{Sht}_T^r], \theta_*^r[\mathrm{Sht}_T^r] \rangle_{\mathrm{Sht}_G^{\prime r}}.$$

In particular, when $r = 0$, the formula for $L^{(0)}(\eta, 0)$ recovers the class number formula

$$L(\eta, 0) = \frac{\#\mathrm{Jac}(X')(k)}{\#\mathrm{Jac}(X)(k)}.$$

4.4.1   Strategy of Proof of Theorem 4.32

In what follows in §4.4, we explain some aspects of the proof of Theorem 4.32. The strategy of proof is to compare two relative trace formulae, one which relates to the intersection number.

Geometry of the moduli spaces of shtuka. In §4.4.2, we give a discussion of the moduli spaces of shtuka. These moduli spaces were constructed by Drinfeld [17] and Varshavsky [53] and their theory parallels that of Shimura varieties in the function field setting. We adapt these constructions to define the notion of Heegner-Drinfeld cycles which play the analogous role to Heegner points on certain Shimura varieties. This is where the étale double cover $\nu : X' \to X$ plays a critical role and new stacks of shtuka constructed on tori will be considered. The section concludes with the definition of the intersection number $\mathbb{I}_r(f)$ in Definition 4.44 which appears on the right-hand side of the equation expressed in Theorem 4.32.

Relative trace formulae. In §4.4.3, we introduce the two relative trace formulae to be compared–a "geometric" one constructed on the intersection number $\mathbb{I}_r(f)$ and an analytic one constructed on an analytic quantity $\mathbb{J}_r(f)$, which should be interpreted as a "naive" intersection number. To delve into this theory, we first introduce the Eisenstein ideal which will play a crucial role in selecting "test functions" to formulate the comparison between the two trace formulae. The Eisenstein ideal also helps simplify the spectral expansion of the relative trace formula constructed on $\mathbb{J}_r(f)$ and allows one to make the comparison with special values of L-functions.

After giving the analytic formulation of the analytic relative trace formula constructed on $\mathbb{J}_r(f)$ and forming the relation with special values of L-functions, we discuss the two ways to express the spectral expansion of the intersection number $\mathbb{I}_r(f)$. These are given in Theorem 4.47 and Theorem 4.48; we merely remark here that the first is a spectral expansion on cycles while the second is on cohomology classes. It is needless to say that these agree.

We conclude with a hollistic overview of the geometric arguments in the paper. We reinterpret the orbital expansion of $\mathbb{J}_r(f)$ using the Lefschetz trace formula as a certain point count on a newly constructed moduli space of orbital integrals. We then introduce the moduli spaces which allow us to formulate the analogous "orbital expansion" of $\mathbb{I}_r(f)$; these moduli spaces should be viewed as a stack of "naive" shtuka, and the expansion is likewise obtained as an application of the Lefschetz trace formula. To conclude the argument makes use of a deep theorem in algebraic geometry called the *perverse continuation principle*. This establishes the key identity which compares each term in the orbital expansions of $\mathbb{J}_r(f)$ and $\mathbb{I}_r(f)$, now seen as weighted counts by the Lefschetz trace formula.

4.4.2 The Moduli Stack of Shtuka

The moduli spaces of shtuka are *stacks* and not schemes. There are several resources at this point on the yoga of stacks at the reader's disposal (the Stacks project, for example). Most of the stacks we consider will be Deligne-Mumford (DM) stacks. DM stacks are particularly nice to work with as they always have a scheme atlas, and much of their theory can be developed on this atlas. The reader may consult [31] for a working introduction to the theory of DM stacks.

The stack of vector bundles, Bun_n^μ. The prototype to the moduli stack of shtuka is the classifying space Bun_n, which attaches to a k-scheme S the groupoid of vector bundles $\mathcal{E}$ of rank n on $X \times S$. The stack Bun_n is a smooth, algebraic stack locally of finite type over k of dimension $n^2(g(X)-1)$, and $\pi_0(\mathrm{Bun}_n) = \mathbb{Z}$.

Example 4.34 (The Picard stack). When $n = 1$, the moduli problem satisfied by Bun_n is the same as that of the already familiar Picard stack, Pic_X. The coarse moduli space of Pic_X is the Jacobian of X, Jac_X, and so there exists a degree-preserving map $\mathrm{Bun}_1 \simeq \mathrm{Pic}_X \to \mathrm{Jac}_X$. If, moreover, X has a k-rational point x, we can construct the isomorphism $\mathrm{Pic}_X \xrightarrow{\sim} \mathrm{Jac}_X \times B\mathbb{G}_m$, where the map $\mathrm{Pic}_X \to B\mathbb{G}_m$ is obtained by restricting the universal line bundle on $X \times \mathrm{Pic}_X$ to $\{x\} \times \mathrm{Pic}_X$. This isomorphism shows that the dimension of Pic_X is $g(X) - 1$.

The k-points of Bun_n can be described by the *Weil uniformization*, a description which is analogous to the complex uniformization of Shimura varieties. Specifically, the Weil uniformization provides the following (canonical) isomorphism of groupoids:

$$\mathrm{GL}_n(F) \backslash \mathrm{GL}_n(\mathbb{A}) / \mathrm{GL}_n(\mathbb{O}) \xrightarrow{\sim} \mathrm{Bun}_n(k).$$

Using the Weil uniformization, we can put level structures on $\mathrm{Bun}_n(k)$ classically as follows. Given an effective divisor $D = \sum d_x \cdot x$ on X, define $K_D = \ker(\mathrm{GL}_n(\mathbb{O}) \to \prod_{x \in |X|} \mathrm{GL}_n(\mathcal{O}_x / \varpi_x^{d_x}))$. The *level D structure* on $\mathrm{Bun}_n(k)$ is the double quotient

$$\mathrm{GL}_n(F) \backslash \mathrm{GL}_n(\mathbb{A}) / K_D \simeq \{(\mathcal{E}, \alpha) \mid \mathcal{E} \in \mathrm{Bun}_n(k), \ \alpha : \mathcal{E}|_D \simeq \mathcal{O}_D^{\oplus n}\}$$

where the isomorphism is induced by the Weil uniformization. More generally, a level D structure on Bun_n is the substack of those $\mathcal{E} \in \mathrm{Bun}_n$ satisfying $\mathcal{E}|_{D \times S} \simeq \mathcal{O}_{D \times S}^{\oplus n}$. We denote this substack by $\mathrm{Bun}_{n,D}$.

The Hecke stack, Hk_n^μ. The Hecke stack plays the philosophical role of the ambient space in which the moduli space of GL_n-shtuka is seen as a "closed locus." Let $r \geq 0$ and $\mu = (\mu_1, \ldots, \mu_r)$ be a sequence of dominant coweights of

GL_n such that μ_i is either $\mu_+ = (1, 0, \ldots, 0)$ or $\mu_- = (0, \ldots, 0, 1)$. We introduce the Hecke stack by the following:

Definition 4.35. The Hecke stack, Hk_n^μ, is the groupoid on k-schemes S classifying the following data:

1. a sequence $\underline{\mathcal{E}} = (\mathcal{E}_0, \ldots, \mathcal{E}_r)$ of rank n vector bundles on $X \times S$,
2. a sequence $\underline{x} = (x_1, \ldots, x_r)$ of morphisms $x_i : S \to X$ with graphs $\Gamma_{x_i} \subseteq X \times S$, and
3. a sequence $\underline{f} = (f_1, \ldots, f_r)$ of maps $f_i : \mathcal{E}_{i-1} \to \mathcal{E}_i$ inducing the isomorphism

$$f_i : \mathcal{E}_{i-1}|_{X \times S \setminus \Gamma_{x_i}} \xrightarrow{\sim} \mathcal{E}_i|_{X \times S \setminus \Gamma_{x_i}}$$

and satisfying the extension property that if $\mu_i = \mu_+$, then f_i extends on all of $X \times S$ to an injection of vector bundles $\mathcal{E}_{i-1} \hookrightarrow \mathcal{E}_i$, and that if $\mu_i = \mu_-$, then f_i^{-1} extends to $\mathcal{E}_i \hookrightarrow \mathcal{E}_{i-1}$. Moreover, the subquotient bundle constructed on the cokernel of f_i or f_i^{-1} for $\mu_i = \mu_+$ or $\mu_i = \mu_-$, respectively, is isomorphic to $\Gamma_{x_i,*}\mathcal{L}_i$ for some line bundle $\mathcal{L}_i$ on S.

For each $i = 0, \ldots, r$, we can construct projection maps to Bun_n by

$$p_i : \mathrm{Hk}_n^\mu \to \mathrm{Bun}_n$$
$$(\underline{\mathcal{E}}, \underline{x}, \underline{f}) \mapsto \mathcal{E}_i.$$

We can also construct a projection to X^r by

$$p_X : \mathrm{Hk}_n^\mu \to X^r$$
$$(\underline{\mathcal{E}}, \underline{x}, \underline{f}) \mapsto \underline{x}.$$

The product of these projections $p_i \times p_X$ for any $i = 0, \ldots, r$ is representable by a proper smooth morphism of relative dimension $r(n-1)$ whose fibers are iterated $\mathbb{P}^{n-1}$-bundles. This shows in particular that Hk_n^μ is isomorphic to Bun_n when $r = 0$.

We can put level structures on Hk_n^μ in such a way as to extend the level D structures already defined on Bun_n. Let $D \subseteq X \times S$ be an effective divisor satisfying $D \cap \{x_1, \ldots, x_r\} = \emptyset$ for a sequence of morphisms $\underline{x} = (x_1, \ldots, x_r) : S \to X$. Then a *level D structure* on Hk_n^μ is the substack of triples $(\underline{\mathcal{E}}, \underline{x}, \underline{f}) \in \mathrm{Hk}_n^\mu$ satisfying the isomorphism $\mathcal{E}_0|_{D \times S} \xrightarrow{\sim} \mathcal{O}_{D \times S}^{\oplus n}$. We denote this substack by $\mathrm{Hk}_{n,D}^\mu$. Under p_i for $i = 1, \ldots, r$, the stack $\mathrm{Hk}_{n,D}^\mu$ projects to $\mathrm{Bun}_{n,D}$, and under p_X it projects to the open curve $X^r \setminus \{x_1, \ldots, x_r\}$.

Moduli of shtuka for GL_n, Sht_n^μ. A shtuka of type μ and rank n is, roughly speaking, a "Hecke modification" together with a Frobenius structure. More precisely, we have the following:

Definition 4.36. The moduli stack of shtuka for GL_n, Sht_n^μ, is the groupoid on k-schemes S classifying triples $(\mathcal{E}, \underline{x}, \underline{f}) \in Hk_n^\mu$ which satisfy the isomorphism $\mathcal{E}_r \simeq {}^\sigma \mathcal{E}_0 := (Id_X \times Frob_S)^* \mathcal{E}_0$.

This definition is nicely summarized by the fiber diagram

$$
\begin{array}{ccc}
Sht_n^\mu & \longrightarrow & Hk_n^\mu \\
\downarrow & & \downarrow{\scriptstyle p_0 \times p_r} \\
Bun_n & \xrightarrow{\ Id \times Frob\ } & Bun_n \times Bun_n.
\end{array}
\tag{4.63}
$$

Example 4.37. Sht_n^μ relate to familiar objects in the following special cases.

1. When $n = 1$, we have $\mathcal{E}_i = \mathcal{E}_{i-1} \otimes \mathcal{O}(x_i)$ and hence $Hk_1^\mu \simeq Pic_X \times X^r$. Thus, the "shtuka" condition for a triple $(\mathcal{E}, \underline{x}, \underline{f}) \in Hk_1^\mu$ translates to

$$
{}^\sigma \mathcal{E}_0 \otimes \mathcal{E}_0^{-1} \simeq \mathcal{O}(\sum \mu_i x_i).
$$

   This is a familiar object classically known as a "Lang torsor" as it is a fiber of the Lang isogeny $Pic_X \xrightarrow{\sigma - 1} Pic_X$ and hence a torsor for $Pic_X(k)$.
2. For $r = 0$, $Sht_n^\mu(S)$ parametrizes the pair $(\mathcal{E}, \alpha : \mathcal{E} \xrightarrow{\sim} {}^\sigma\mathcal{E})$ where $\mathcal{E}$ is a vector bundle of rank n on $X \times S$. This gives an isomorphism $Sht_n^\mu(k) \simeq Bun_n(k)$.

By Drinfeld [17] in the case $r = 2$ and Varshavsky [53] in general, Sht_n^μ is shown to be a smooth Deligne-Mumford stack which is locally of finite type over k. The key to show Sht_n^μ is locally of finite type is to construct a "cover" of Sht_n^μ by DM stacks $Sht_{n,D,d,m}^\mu$ which are of finite type by their definition. This is done by introducing both level structures and stability conditions on the moduli stack of shtuka.

A *level D structure* on Sht_n^μ is induced from a level D structure on Hk_n^μ. Specifically, it is the substack of triples classified by $Hk_{n,D}^\mu$ satisfying the commutativity condition

$$
\begin{array}{ccc}
\mathcal{E}_0|_{D \times S} & \xrightarrow{\ \sim\ } & \mathcal{O}_{D \times S}^{\oplus n} \\
\downarrow{\scriptstyle \sim} & & \| \\
{}^\sigma \mathcal{E}_0|_{D \times S} & \xrightarrow{\ \sim\ } & \mathcal{O}_{D \times S}^{\oplus n, \sigma}.
\end{array}
$$

A stability condition on Sht_n^μ is defined with respect to the following measure on elements $\mathcal{E}$ of Bun_n:

$$
M(\mathcal{E}) := \max\{\deg \mathcal{L} | \mathcal{L} \hookrightarrow \mathcal{E}\}.
$$

A collection of vector bundles is *stable* if all the vector bundles in the collection are in the same component of Bun_n and $M(\mathcal{E})$ is bounded for each vector bundle $\mathcal{E}$ in the collection.

Definition 4.38. The stack $\mathrm{Sht}^{\mu}_{n,D,d,m}$ is the groupoid on k-schemes S classifying triples $(\mathcal{E}, \underline{x}, f) \in \mathrm{Sht}^{\mu}_n$ satisfying the conditions

1. $(\mathcal{E}, \underline{x}, f)$ has a level D structure for a fixed effective divisor $D \subseteq X$,
2. $\deg(\det \mathcal{E}_i) = d$ for a fixed non-negative integer d, and
3. $M(\mathcal{E}_0) \leq m$ for a fixed non-negative integer m.

Moduli of shtuka for PGL_n. Let $G = \mathrm{PGL}_n = \mathrm{GL}_n / \mathbb{G}_m$. We construct the moduli space of G-shtuka formally as a stack quotient of Sht^{μ}_n. This construction is already found for the stack Bun_G of G-torsors on X, which is canonically isomorphic to the stack quotient $\mathrm{Bun}_n / \mathrm{Pic}_X$. The action here of Pic_X on Bun_n is obtained by forming a tensor product of a representative element of a class in Bun_n with a representative element of a class in Pic_X. It is easy to see that the class of the resulting tensor object is independent of the choices made.

The action of Pic_X on Bun_n lifts to Hk^{μ}_n by

$$(\mathcal{E}, \underline{x}, f) \mapsto (\mathcal{E} \otimes \mathcal{L}, \underline{x}, f \otimes \mathrm{Id}),$$

where $\mathcal{L}$ is a line bundle representing a class in Pic_X. The action of Pic_X on Hk^{μ}_n restricts to Sht^{μ}_n whenever $\mathcal{L} \simeq {}^{\sigma}\mathcal{L}$, which is satisfied by elements of $\mathrm{Pic}(k)$. This gives an action of $\mathrm{Pic}_X(k)$ on Sht^{μ}_n compatible with the inclusion $\mathrm{Sht}^{\mu}_n \hookrightarrow \mathrm{Hk}^{\mu}_n$. As a result, the commutative diagram

$$\begin{array}{ccc} \mathrm{Pic}_X(k) & \longrightarrow & \mathrm{Pic}_X \\ \downarrow & & \downarrow{\scriptstyle p_0 \times p_0} \\ \mathrm{Pic}_X & \xrightarrow{\mathrm{Id} \times \mathrm{Frob}} & \mathrm{Pic}_X \times \mathrm{Pic}_X \end{array} \qquad (4.64)$$

gives a compatible action on each respective object in (4.63).

Definition 4.39. The moduli stack of shtuka for $G = \mathrm{PGL}_n$, Sht^{μ}_G, is the quotient of the fiber diagram (4.63) by the action of the diagram (4.64). In particular, it is the stack defined by the stack quotient

$$\mathrm{Sht}^{\mu}_G = \mathrm{Sht}^{\mu}_n / \mathrm{Pic}_X(k).$$

When $n = 2$, Drinfeld [17] demonstrates the following phenomenon. For any pair of r-tuples of signs μ, μ', there is a canonical isomorphism

$$\mathrm{Sht}^{\mu}_G \xrightarrow{\sim} \mathrm{Sht}^{\mu'}_G.$$

Indeed, if $(\mathcal{E}, \underline{x}, f) \in \mathrm{Sht}^{\mu}_G$ for any μ, we can transform it to an element on $\mathrm{Sht}^{\mu'}_G$ for $\mu' = (\mu_+, \ldots, \mu_+)$ by constructing $\mathcal{E}'_i = \mathcal{E}_i \otimes \mathcal{O}_{X \times S}(D_i)$ where $D_i =$

$\sum_{\substack{1 \le j \le i \\ \mu_j = \mu_-}} \Gamma_{x_i}$. Thus, for $n = 2$ and r even, we may fix a choice of μ with the following sign convention:

$$\mu_i = \begin{cases} \mu_+ & \text{if } 1 \le i \le r/2, \\ \mu_- & \text{if } r/2 < i \le r. \end{cases}$$

We call such a μ a *balanced r-tuple*, or more simply *balanced*, and denote the stack of shtuka for this choice of μ by Sht_n^r or Sht_G^r as in §4.1.4.

Hecke correspondences on PGL_2-shtuka. We now fix $G = \mathrm{PGL}_2$ and wish to describe the action given by Hecke correspondences on Sht_G^r. The rational Chow group of proper cycles $\mathrm{Ch}_{c,*}(\mathrm{Sht}_G^r)_{\mathbb{Q}}$ is defined in Appendix A.1 of [61], along with the group ${}_c\mathrm{Ch}_{2r}(\mathrm{Sht}_G^r \times \mathrm{Sht}_G^r)_{\mathbb{Q}}$ of rational cycles of dimension $2r$ which are proper over the first factor. By A.1.6, [61], the group ${}_c\mathrm{Ch}_{2r}(\mathrm{Sht}_G^r \times \mathrm{Sht}_G^r)_{\mathbb{Q}}$ acts on $\mathrm{Ch}_{c,*}(\mathrm{Sht}_G^r)_{\mathbb{Q}}$.

Recall first the classical definition of the unramified Hecke algebra, which we specialize to the reductive group $G = \mathrm{PGL}_2$ below. We write by $K = \prod_{x \in |X|} K_x \subseteq G(\mathbb{A})$ the compact subgroup where the local factors are defined to be the maximal compact subgroup $K_x = G(\mathcal{O}_x) \subseteq G(F_x)$.

Definition 4.40. The local unramified Hecke algebra $\mathcal{H}_x$ at $x \in |X|$ for $G = \mathrm{PGL}_2$ is the convolution algebra

$$\mathcal{H}_x = C_c^\infty(K_x \backslash G(F_x)/K_x, \mathbb{Q}).$$

This is canonically identified with the polynomial algebra $\mathbb{Q}[h_x]$ and has a $\mathbb{Q}$-basis given by the functions h_{dx}, which are defined for all $d \in \mathbb{Z}_{\ge 0}$ to be the characteristic functions on the double coset

$$K_x \begin{pmatrix} \varpi_x^d & \\ & 1 \end{pmatrix} K_x$$

where $\varpi_x \in \mathcal{O}_x$ is any choice of a uniformizing element.

The global unramified Hecke algebra $\mathcal{H}$ for $G = \mathrm{PGL}_2$ is the convolution algebra

$$\mathcal{H} := C_c^\infty(K \backslash G(\mathbb{A})/K, \mathbb{Q}) = \bigotimes_{x \in |X|} \mathcal{H}_x.$$

For each effective divisor $D = \sum_{x \in |X|} d_x x \subseteq X$, we can define an element $h_D = \bigotimes_{x \in |X|} h_{d_x x} \in \mathcal{H}$. The h_D form a $\mathbb{Q}$-basis for $\mathcal{H}$ and are called Hecke correspondences.

For each Hecke correspondence h_D we shall define a self-correspondence $\mathrm{Sht}_G^r(h_D)$ of Sht_G^r over X^r in the following way.

Definition 4.41. The stack of Hecke correspondences for an effective divisor $D \subseteq X$, $\mathrm{Sht}_2^r(h_D)$, is the groupoid on k-schemes S classifying the data:

1. pairs of objects $(\mathcal{E}, \underline{x}, f)$ and $(\mathcal{E}', \underline{x}, f')$ of Sht_2^r with the same collection of points $\underline{x} = (x_1, \ldots, x_r)$ in $X(S)$,
2. a morphism $\phi : (\mathcal{E}, \underline{x}, f) \to (\mathcal{E}', \underline{x}, f')$ satisfying that for each $i = 0, \ldots, r$ the morphism $\phi_i : \mathcal{E}_i \hookrightarrow \mathcal{E}'_i$ gives an embedding of coherent sheaves such that $\det(\phi_i) : \det\mathcal{E}_i \hookrightarrow \det\mathcal{E}'_i$ has divisor $D \times S \subseteq X \times S$, and further that the following diagram is commutative

$$
\begin{array}{ccccccccc}
\mathcal{E}_0 & \xrightarrow{f_1} & \mathcal{E}_1 & \xrightarrow{f_2} & \cdots & \xrightarrow{f_r} & \mathcal{E}_r & \xrightarrow{\cong} & \mathcal{E}_0^\sigma \\
\downarrow{\phi_0} & & \downarrow{\phi_1} & & & & \downarrow{\phi_r} & & \downarrow{\phi_0^\sigma} \\
\mathcal{E}'_0 & \xrightarrow{f'_1} & \mathcal{E}'_1 & \xrightarrow{f'_2} & \cdots & \xrightarrow{f'_r} & \mathcal{E}'_r & \xrightarrow{\cong} & (\mathcal{E}'_0)^\sigma.
\end{array}
$$

There is a natural action of $\mathrm{Pic}_X(k)$ on $\mathrm{Sht}_2^r(h_D)$ given by tensor product on each $\mathcal{E}_i$ and $\mathcal{E}'_i$. This allows us to define for $G = \mathrm{PGL}_2$ the stack of Hecke correspondences for Sht_G^r by

$$
\mathrm{Sht}_G^r(h_D) := \mathrm{Sht}_2^r(h_D)/\mathrm{Pic}_X(k).
$$

This stack sits in a diagram

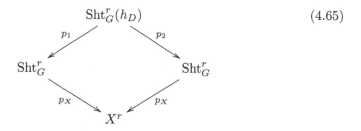

(4.65)

where the maps p_1 and p_2 project onto the source and target, respectively, of the classified morphism ϕ. One can show that the projections p_i for $i = 1, 2$ are representable and proper and are mutually transverse. Moreover, the geometric fibers of the compositions $p_i \circ p_X : \mathrm{Sht}_G^r(h_D) \to X^r$ for $i = 1, 2$ each have dimension r. This shows that $\dim \mathrm{Sht}_G^r(h_D) = 2r$.

By these facts, it makes sense to pushforward the fundamental cycle of $\mathrm{Sht}_G^r(h_D)$ along the proper map $(p_1, p_2) : \mathrm{Sht}_G^r(h_D) \to \mathrm{Sht}_G^r \times \mathrm{Sht}_G^r$. This induces a $\mathbb{Q}$-linear map

$$
H : \mathcal{H} \to {}_c\mathrm{Ch}_{2r}(\mathrm{Sht}_G^r \times \mathrm{Sht}_G^r)_\mathbb{Q}
$$
$$
h_D \mapsto (p_1, p_2)_*[\mathrm{Sht}_G^r(h_D)].
$$

By Proposition 5.10 [61] H is a ring homomorphism, and so by A.1.6 [61] defines an action on $\mathrm{Ch}_{c,r}(\mathrm{Sht}_G^r)$ for each $f \in \mathcal{H}$ on $\mathrm{Ch}_{c,*}(\mathrm{Sht}_G^r)_\mathbb{Q}$. We denote this action by

$$f * (-) : \mathrm{Ch}_{c,*}(\mathrm{Sht}_G^r)_{\mathbb{Q}} \to \mathrm{Ch}_{c,*}(\mathrm{Sht}_G^r)_{\mathbb{Q}}.$$

By Lemma 5.12 [61], this action is self-adjoint with respect to the intersection pairing on $\mathrm{Ch}_{c,*}(\mathrm{Sht}_G^r)_{\mathbb{Q}}$.

Heegner-Drinfeld cycles. The stack of *Heegner-Drinfeld cycles* will play the analogous role to that of Heegner points or Heegner divisors on Shimura varieties in the classical Gross-Zagier formulae. We continue to fix $G = \mathrm{PGL}_2$ and let μ denote an r-tuple of signs, specifying that r is even and μ is balanced only when needed.

We first consider the following mild variants of the previous constructions. To geometrize the notion of quadratic points on a Shimura variety, we use the double cover $\nu : X' \to X$ to construct the stack

$$\mathrm{Sht}_G^{\prime\mu} := \mathrm{Sht}_G^{\mu} \times_{X^r} X^{\prime r}.$$

When r is even and μ is balanced, its stack of Hecke correspondences is also obtained by base change as $\mathrm{Sht}_G^{\prime r}(h_D) := \mathrm{Sht}_G^r(h_D) \times_{X^r} X^{\prime r}$ and is equipped with the natural projection maps $p_1', p_2' : \mathrm{Sht}_G^{\prime r}(h_D) \to \mathrm{Sht}_G^{\prime r}$. This induces the ring homomorphism

$$H' : \mathcal{H} \to {}_c\mathrm{Ch}_{2r}(\mathrm{Sht}_G^{\prime r} \times \mathrm{Sht}_G^{\prime r})_{\mathbb{Q}}$$
$$h_D \mapsto (p_1', p_2')_*[\mathrm{Sht}_G^{\prime r}(h_D)]$$

and thus an action of $f \in \mathcal{H}$ on $\mathrm{Ch}_{c,*}(\mathrm{Sht}_G^{\prime r})_{\mathbb{Q}}$. We continue to denote this action by $f * (-)$.

For the actual Heegner-Drinfeld cycles on $\mathrm{Sht}_G^{\prime\mu}$ we need an analogue of CM structures to shtuka. We define two tori: the two-dimensional torus $\widetilde{T} := \mathrm{Res}_{X'/X} \mathbb{G}_m$ and the one-dimensional torus $T := \widetilde{T}/\mathbb{G}_m$. There are natural inclusions $\widetilde{T} \hookrightarrow \mathrm{GL}_2$ and $T \hookrightarrow G$ and an identification $\mathrm{Bun}_T \sim \mathrm{Pic}_{X'}/\mathrm{Pic}_X$.

Definition 4.42. The moduli stack of $\widetilde{T}$-Shtuka, $\mathrm{Sht}_{\widetilde{T}}^{\mu}$, is the groupoid on k-schemes S classifying the following data:

1. a line bundle $\mathcal{L}$ over $X' \times S$,
2. a sequence $\underline{x}' = (x_1', \ldots, x_r')$ of morphisms $x_i' : S \to X'$ with graphs $\Gamma_{x_i'} \subseteq X' \times S$, and
3. an isomorphism

$$\iota : \mathcal{L}\left(\sum_{i=1}^{r} \mathrm{sgn}(\mu_i)\Gamma_{x_i'}\right) \xrightarrow{\sim} {}^{\sigma}\mathcal{L} := (\mathrm{id} \times \mathrm{Fr}_S)^* \mathcal{L}.$$

The stack $\mathrm{Sht}_{\widetilde{T}}^{\mu}$ can be identified with the stack $\mathrm{Sht}_{1,X'}^{\mu}$, which is given by Definition 4.36 after replacing X with X' and specializing to $n = 1$. Thus,

$\mathrm{Sht}^{\mu}_{\widetilde{T}}$ sits in the diagram

$$
\begin{array}{ccc}
\mathrm{Sht}^{\mu}_{\widetilde{T}} & \longrightarrow \mathrm{Hk}^{\mu}_{1,X'} \xrightarrow{\ p_{X'}\ } X'^{r} \\
\downarrow & \qquad\downarrow{\scriptstyle p_0 \times p_r} \\
\mathrm{Pic}_{X'} & \xrightarrow{\ \mathrm{Id} \times \mathrm{Frob}\ } \mathrm{Pic}_{X'} \times \mathrm{Pic}_{X'}
\end{array}
$$

in which $\mathrm{Hk}^{r}_{1,X'}$ is given by replacing X with X' in Definition 4.35. We let $\pi^{\mu}_{\widetilde{T}} : \mathrm{Sht}^{\mu}_{\widetilde{T}} \to X'^{r}$ be the composition in the top row of the diagram. We can further define a morphism

$$
\widetilde{\theta}^{\mu} : \mathrm{Sht}^{\mu}_{\widetilde{T}} \to \mathrm{Sht}^{\mu}_{2}
$$

by using the identification between $\mathrm{Sht}^{\mu}_{\widetilde{T}}$ with $\mathrm{Sht}^{\mu}_{1,X'}$ and mapping along the pushforward $(\mathcal{L}, \underline{x}, \underline{f}) \mapsto (\nu_{*}(\mathcal{L}), \underline{x}, \underline{f} \circ \nu_{*})$. Note that for different choices of μ we can intertwine the maps $\widetilde{\theta}^{\mu}$ and that it respects the projections on the target to X^{r} but not on the source to X'^{r} (see §5.4.6 [61] for a discussion).

Now we pass these constructions to their $\mathrm{Pic}_{X}(k)$-quotients. We begin by defining the stack

$$
\mathrm{Sht}^{\mu}_{T} := \mathrm{Sht}^{\mu}_{\widetilde{T}} / \mathrm{Pic}_{X}(k)
$$

where $\mathrm{Pic}_{X}(k)$ acts on $\mathrm{Sht}^{\mu}_{\widetilde{T}}$ via tensoring along the pushforward $\nu_{*} : \mathrm{Pic}_{X}(k) \to \mathrm{Pic}_{X'}(k)$. The morphisms $\pi^{\mu}_{\widetilde{T}}$ and $\widetilde{\theta}^{\mu}$ are both $\mathrm{Pic}_{X}(k)$-invariant, so we have induced morphisms π^{μ}_{T} and $\overline{\theta}^{\mu}$. These fit into the diagram

$$
\begin{array}{ccc}
\mathrm{Sht}^{\mu}_{T} & \xrightarrow{\ \overline{\theta}^{\mu}\ } & \mathrm{Sht}^{r}_{G} \\
{\scriptstyle \pi^{\mu}_{T}}\downarrow & & \downarrow{\scriptstyle \pi_{G}} \\
X'^{r} & \xrightarrow{\ \nu^{r}\ } & X^{r}
\end{array}
$$

which induces the morphism $\theta^{\mu} : \mathrm{Sht}^{\mu}_{T} \to \mathrm{Sht}'^{\mu}_{G}$ by factoring $\overline{\theta}^{\mu}$ through the fiber product. It can be shown that $\theta^{\mu}_{*}[\mathrm{Sht}^{\mu}_{T}]$ defines a proper cycle on Sht'^{μ}_{G} of dimension r.

Definition 4.43. The Heegner-Drinfeld cycle of type μ is the element

$$
\theta^{\mu}_{*}[\mathrm{Sht}^{\mu}_{T}] \in \mathrm{Ch}_{c,r}(\mathrm{Sht}'^{\mu}_{G})_{\mathbb{Q}}.
$$

Now fixing r to be even and μ to be balanced, recall that we have an action of $\mathcal{H}$ on $\mathrm{Ch}_{c,r}(\mathrm{Sht}'^{r}_{G})_{\mathbb{Q}}$. Both $\theta^{r}_{*}[\mathrm{Sht}^{r}_{T}]$ and $f * \theta^{r}_{*}[\mathrm{Sht}^{r}_{T}]$ for any $f \in \mathcal{H}$ define proper cycle classes in $\mathrm{Ch}_{c,r}(\mathrm{Sht}'^{r}_{G})_{\mathbb{Q}}$ of complementary dimension. Thus, we can formulate the following definition.

Definition 4.44. Let $f \in \mathcal{H}$ be an unramified Hecke function. Attached to f is the intersection number

$$\mathbb{I}_r(f) := \langle \theta_*^r[\mathrm{Sht}_T^r], f * \theta_*^r[\mathrm{Sht}_T^r] \rangle_{\mathrm{Sht}_G'^r} \in \mathbb{Q}.$$

4.4.3 The Relative Trace Formulae

The basic strategy to prove Theorem 4.32 is to compare two relative trace formulae. A relative trace formula (abbreviated as RTF) is an equality between a spectral expansion and an orbital integral expansion of a numeric quantity. The information obtained from the spectral expansion is used to evaluate a specific quantity—in this case either an intersection number or the special value of an L-function—while the orbital expansion allows one to compare the two relative trace formulae term-wise. Here, what we term the "geometric" RTF relates to the intersection numbers and the "analytic" RTF which relates to the special values of the L-function; these quantities are in analogy to the two sides of the equality in the classical *Gross-Zagier* formula. This strategy of comparison is summarized by the following diagram, where the top row corresponds to the analytic RTF and the bottom row corresponds to the geometric RTF:

Analytic RTF: $\sum_{u \in \mathbb{P}^1(F)-\{1\}} \mathbb{J}_r(u, f) =\!\!=\!\!= \mathbb{J}_r(f) =\!\!=\!\!= \sum_\pi \mathbb{J}_r(\pi, f)$

$$\Big\downarrow \sim \qquad\qquad \Big\downarrow \sim \qquad\qquad \Big\downarrow \,{\cdot}{\cdot}$$

Geometric RTF: $\sum_{u \in \mathbb{P}^1(F)-\{1\}} \mathbb{I}_r(u, f) =\!\!=\!\!= \mathbb{I}_r(f) =\!\!=\!\!= \sum_{\mathfrak{m}} \mathbb{I}_r(\mathfrak{m}, f).$

In this diagram, $\sim$ means equality after dividing the top row by the quantity $(\log q)^r$. The quantities in this diagram will be defined in the paragraphs below.

The Eisenstein ideal. Let $G = \mathrm{PGL}_2$ be defined over F and $A \subseteq G$ be the diagonal subtorus of G. Recall that $\mathcal{H} = \otimes_{x \in |X|} \mathcal{H}_x$ is the unramified Hecke algebra for G (given in Definition 4.40) which has a basis given by the Hecke operators $\{h_D\}$ indexed by effective divisors D on X. Consider also $\mathcal{H}_A = \otimes_{x \in |X|} \mathcal{H}_{A,x}$ the unramified Hecke algebra for A, for which we have the identification $\mathcal{H}_{A,x} = \mathbb{Q}[F_x^\times / \mathcal{O}_x^\times] \simeq \mathbb{Q}[t_x, t_x^{-1}]$ where t_x stands for the characteristic function of $\varpi_x^{-1} \mathcal{O}_x^\times$. We also have the identification $\mathcal{H}_A = \mathbb{Q}[\mathbb{A}^\times / \mathbb{O}^\times] \simeq \mathbb{Q}[\mathrm{Div}(X)]$.

The (normalized) local Satake transform is the map

$$\mathrm{Sat}_x : \mathcal{H}_x \to \mathcal{H}_{A,x}$$
$$h_x \mapsto t_x + q_x t_x^{-1}$$

where $q_x = \#k_x$. Denote by ι_x the involution on $\mathcal{H}_{A,x}$ that sends t_x to $q_x t_x^{-1}$. Then Sat_x identifies $\mathcal{H}_x$ with the subring of ι_x-invariants of $\mathcal{H}_{A,x}$. The global

Satake transform is the tensor product

$$\text{Sat} = \bigotimes_{x \in |X|} \text{Sat}_x : \mathcal{H} \to \mathcal{H}_A.$$

Consider the composition

$$\mathcal{H} \xrightarrow{\text{Sat}} \mathcal{H}_A \simeq \mathbb{Q}[\text{Div}(X)] \twoheadrightarrow \mathbb{Q}[\text{Pic}_X(k)].$$

The image of this composition lands in the ι_{Pic}-invariants of $\mathbb{Q}[\text{Pic}_X]$, where ι_{Pic} is the involution given by

$$\iota_{\text{Pic}}(\mathbf{1}_{\mathcal{L}}) = q^{\deg \mathcal{L}} \mathbf{1}_{\mathcal{L}^{-1}}, \quad \mathcal{L} \in \text{Pic}_X(k)$$

and is identified with the involution $\bigotimes_x \iota_x$ on $\mathcal{H}_A$ under the projection $\mathbb{Q}[\text{Div}(X)] \twoheadrightarrow \mathbb{Q}[\text{Pic}_X(k)]$. This defines the ring homomorphism

$$a_{\text{Eis}} : \mathcal{H} \to \mathbb{Q}[\text{Pic}_X(k)]^{\iota_{\text{Pic}}} =: \mathcal{H}_{\text{Eis}}.$$

Definition 4.45. The Eisenstein ideal $\mathcal{I}_{\text{Eis}} \subseteq \mathcal{H}$ is the kernel of the ring homomorphism a_{Eis}.

The analytic relative trace formula. We define the analytic quantity $\mathbb{J}_r(f)$ as follows. To an $f \in \mathcal{H}$ (or more generally $f \in C_c^\infty(G(\mathbb{A}))$), we associate an automorphic kernel function $\mathbb{K}_f$ on $G(\mathbb{A}) \times G(\mathbb{A})$ by the formula

$$\mathbb{K}_f(g_1, g_2) = \sum_{\gamma \in G(F)} f(g_1^{-1} \gamma g_2).$$

We also define the regularized integral (see §2.2 [61] for a discussion on regularization)

$$\mathbb{J}(f, s) = \int_{[A] \times [A]}^{\text{reg}} \mathbb{K}_f(h_1, h_2) |h_1 h_2|^s \eta(h_2) dh_1 dh_2 \in \mathbb{Q}[q^s, q^{-s}],$$

where for $h = \begin{pmatrix} x & \\ & y \end{pmatrix}$ we define $|h| = |x/y|$, and $\eta : \mathbb{A}^\times \to \{\pm 1\}$ is the quadratic character corresponding to the extension F'/F by class field theory. Informally, this integral is a weighted, naive intersection number on $\text{Bun}_G(k)$ between $\text{Bun}_A(k)$ and its Hecke translation under f. The quantity $\mathbb{J}_r(f)$ is the r-th derivative

$$\mathbb{J}_r(f) = \left(\frac{d}{ds} \right)^r \mathbb{J}(f, s) \Big|_{s=0}.$$

When $r = 0$, the relative trace formula for $\mathbb{J}_r(f)$ was shown by Jacquet, and the general case builds on his approach. The orbital decomposition we want is along the invariant map

$$\text{inv} : G \to \mathbb{P}^1 - \{1\}$$

$$\begin{pmatrix} a & b \\ c & d \end{pmatrix} \mapsto \frac{bc}{ad}.$$

Note that this morphism is $A \times A$-invariant where the action of $A \times A$ on G is given by $(h_1, h_2)g = h_1^{-1}gh_2$, and thus the map factors through the orbit space $A(F) \backslash G(F) / A(F)$. For each $u \in \mathbb{P}^1(F) - \{0, 1, \infty\}$, one can show that $\text{inv}^{-1}(u)$ contains a unique orbit, and otherwise for each $u \in \{0, \infty\}$ the fiber $\text{inv}^{-1}(u)$ is a set of three orbits (see §2.1 [61]).

To any orbit $\gamma \in A(F) \backslash G(F) / A(F)$, we attach the kernel

$$\mathbb{K}_{f,\gamma}(h_1, h_2) = \sum_{\delta \in A(F)\gamma A(F)} f(h_1^{-1}\delta h_2)$$

and define the regularized integral

$$\mathbb{J}(\gamma, f, s) = \int_{[A] \times [A]}^{\text{reg}} \mathbb{K}_{f,\gamma}(h_1, h_2) |h_1 h_2|^s \eta(h_2) dh_1 dh_2.$$

Formally, we obtain the fine orbital decomposition $\mathbb{K}_f(h_1, h_2) = \sum_{\gamma \in A(F) \backslash G(F) / A(F)} \mathbb{K}_{f,\gamma}(h_1, h_2)$ and $\mathbb{J}(f, s) = \sum_{\gamma \in A(F) \backslash G(F) / A(F)} \mathbb{J}(\gamma, f, s)$. In Proposition 2.3, [61], it is shown that these decompositions hold analytically using the appropriate regularization of the integral to define $\mathbb{J}(f, s)$ and $\mathbb{J}(\gamma, f, s)$. To form the comparison between the orbital decomposition factors of the two relative trace formulae, we consider a slightly coarser variant. For $u \in \mathbb{P}^1(F) - \{1\}$, let

$$\mathbb{J}(u, f, s) = \sum_{\substack{\gamma \in A(F) \backslash G(F) / A(F) \\ \text{inv}(\gamma) = u}} \mathbb{J}(\gamma, f, s)$$

and define the quantity

$$\mathbb{J}_r(u, f) = \left(\frac{d}{ds}\right)^r \mathbb{J}(u, f, s)\Big|_{s=0}.$$

Then we have the orbital expansion of $\mathbb{J}_r(f)$ by

$$\mathbb{J}_r(f) = \sum_{u \in \mathbb{P}^1(F) - \{1\}} \mathbb{J}_r(u, f).$$

The spectral decomposition of $\mathbb{J}_r(f)$ is obtained from the decomposition of the kernel function into a cuspidal, special, and Eisenstein part (see §4.2 [61] for definitions):

$$\mathbb{K}_f = \mathbb{K}_{f,\mathrm{cusp}} + \mathbb{K}_{f,\mathrm{sp}} + \mathbb{K}_{f,\mathrm{Eis}}.$$

In general, it is difficult to describe $\mathbb{J}_r(f)$ along this decomposition, but for well-chosen test functions $f \in \mathcal{H}$ it can take a simple form. In particular, it is first shown in Theorem 4.3 of [61] that for $f \in \mathcal{I}_{\mathrm{Eis}}$, the decomposition of the automorphic kernel function simplifies to $\mathbb{K}_f = \mathbb{K}_{f,\mathrm{cusp}} + \mathbb{K}_{f,\mathrm{sp}}$. In Lemma 4.4 [61], it is further shown that only the cuspidal part contributes to the computation of $\mathbb{J}(f, s)$, i.e., for $f \in \mathcal{I}_{\mathrm{Eis}}$, we have the equality

$$\mathbb{J}(f, s) = \int_{[A] \times [A]}^{\mathrm{reg}} \mathbb{K}_{f,\mathrm{cusp}}(h_1, h_2) |h_1 h_2|^2 \eta(h_2) dh_1 dh_2$$

$$= \sum_{\pi} \int_{[A] \times [A]}^{\mathrm{reg}} \mathbb{K}_{f,\pi}(h_1, h_2) |h_1 h_2|^2 \eta(h_2) dh_1 dh_2,$$

where the sum is over all unramified cuspidal automorphic representations π of G. Note that this sum is finite and that for each cuspidal automorphic representation π of G, the π-component of the kernel function is

$$\mathbb{K}_{f,\pi}(g_1, g_2) = \sum_{\phi} \pi(f)\phi(g_1)\overline{\phi(g_2)},$$

where the sum runs over an orthonormal basis $\{\phi\}$ of π. We write the piece corresponding to π in the cuspidal decomposition of $\mathbb{J}(f, s)$ by $\mathbb{J}_\pi(f, s)$, which yields, for each $f \in \mathcal{I}_{\mathrm{Eis}}$, the spectral decomposition formula

$$\mathbb{J}(f, s) = \sum_{\pi} \mathbb{J}_\pi(f, s).$$

Relation to L-functions. Given an everywhere unramified cuspidal automorphic representation π with coefficient field E_π, we let $L(\pi, s)$ denote the standard (complete) L-function associated to π. This L-function can be shown to be a polynomial of degree $4(g-1)$ in the variable $q^{-s-1/2}$ with coefficients in the ring of integers $\mathcal{O}_{E_\pi} \subseteq E_\pi$. Let $\pi_{F'}$ be the base change of π to F' and $L(\pi_{F'}, s)$ be the standard L-function associated to $\pi_{F'}$. This relates to $L(\pi, s)$ by

$$L(\pi_{F'}, s) = L(\pi, s)L(\pi \otimes \eta, s)$$

where we recall that η is the quadratic character associated to the extension F'/F by class field theory. The relation shows that $L(\pi_{F'}, s)$ is a polynomial of degree $8(g-1)$ in the variable q^{-s-1} with coefficients in E_π. It further satisfies the functional equation

$$L(\pi_{F'}, s) = \epsilon(\pi_{F'}, s)L(\pi_{F'}, 1-s)$$

where $\epsilon(\pi_{F'}, s) = q^{-8(g-1)(s-1/2)}$.

Let $L(\pi, \mathrm{Ad}, s)$ be the adjoint L-function of π. Then define the normalized L-function

$$\mathcal{L}(\pi_{F'}, s) = \epsilon(\pi_{F'}, s)^{-1/2} \frac{L(\pi_{F'}, s)}{L(\pi, \mathrm{Ad}, 1)}.$$

This has the functional equation

$$\mathcal{L}(\pi_{F'}, s) = \mathcal{L}(\pi_{F'}, 1 - s).$$

By the functional equation, we can see that if r is odd then the central value of the r-th derivative $\mathcal{L}^{(r)}(\pi_{F'}, 1/2) = 0$. Otherwise, we can show

$$\mathcal{L}^{(r)}(\pi_{F'}, 1/2) \in E_\pi \cdot (\log q)^r.$$

Theorem 4.46 (Theorem 4.7, [61]). *Let $f \in \mathcal{I}_{Eis}$. Then we have an equality*

$$\mathbb{J}(f, s) = \frac{1}{2}|\omega_X| \sum_\pi \mathcal{L}(\pi_{F'}, s + 1/2)\lambda_\pi(f),$$

where the sum runs over all irreducible cuspidal automorphic $\mathcal{H}$-modules π.

The proof of this theorem reduces to the local identity

$$\mathbb{J}_\pi(f, s) = \frac{1}{2}|\omega_X| \mathcal{L}(\pi_{F'}, s + 1/2)\lambda_\pi(f),$$

for each unramified cuspidal automorphic representation π of G appearing in the sum, which is finite. This identity may be viewed as a "Waldspurger formula" in the following sense.

First, the distribution $\mathbb{J}_\pi(f, s)$ can be written as a product of period integrals

$$\mathbb{J}_\pi(f, s) = \sum_\phi \frac{\mathcal{P}_1(\pi(f)\phi, s)\mathcal{P}_\eta(\overline{\phi}, s)}{\langle \phi, \phi \rangle_{\mathrm{Pet}}}$$

where the sum runs over an orthogonal basis $\{\phi\}$ of π, and the pairing in the denominator is induced from the Petersson inner product. Here, the (A, χ)-period integral for $\phi \in \pi$ and for a character $\chi : F^\times \backslash \mathbb{A}^\times \to \mathbb{C}^\times$ is defined as

$$\mathcal{P}_\chi(\phi, s) := \int_{[A]} \phi(h)\chi(h)|h|^s dh.$$

Then for $\phi \in \pi$, there is a "Waldspurger relation" (equation (4.16) in the proof of Proposition 4.5 in [61])

$$\frac{\mathcal{P}_1(\phi, s)\mathcal{P}_\eta(\overline{\phi}, s)}{\langle \phi, \phi \rangle_{\mathrm{Pet}}} = |\omega_X|^{-1}\frac{L(\pi_{F^\times}, s + 1/2)}{2L(\pi, \mathrm{Ad}, 1)} \prod_{x \in |X|} \xi_{x, \psi_x}(\eta_x, s),$$

where the local terms $\xi_{x,\psi_x}(\eta_x, s)$ relate to the local Whittaker model of π at $x \in |X|$ constructed for an appropriately chosen additive character $\psi: \mathbb{A} \to \mathbb{C}^\times$. By Lemma 4.6, [61], which gives a relation between a factor $\xi_{x,\psi_x}(\eta_x, s)$ and certain twists, one concludes the theorem by explicitly computing the product in the above formula.

Spectral decomposition of intersection numbers. In Definition 4.44 we defined the intersection number

$$\mathbb{I}_r(f) := \langle \theta_*^r[\mathrm{Sht}_T^r], f * \theta_*^r[\mathrm{Sht}_T^r] \rangle_{\mathrm{Sht}_G^{\prime r}} \in \mathbb{Q}, \quad f \in \mathcal{H}.$$

There are two perspectives given in [61] to give a spectral decomposition of this quantity. The first is to view the Heegner-Drinfeld cycle as an element in the Chow group $\mathrm{Ch}_{c,r}(\mathrm{Sht}_G^{\prime r})_{\mathbb{Q}}$ on which the Hecke algebra $\mathcal{H}$ acts as correspondences. In this perspective, we let $\widetilde{W} \subseteq \mathrm{Ch}_{c,r}(\mathrm{Sht}_G^{\prime r})_{\mathbb{Q}}$ be the sub $\mathcal{H}$-module generated by $\theta_*[\mathrm{Sht}_T^\mu]$ and $\widetilde{W}_0$ be the submodule

$$\widetilde{W}_0 = \left\{ x \in \widetilde{W} : \langle z, z' \rangle_{\mathrm{Sht}_G^{\prime r}} = 0, \text{ for all } z' \in \widetilde{W} \right\}.$$

Then the intersection pairing factors through and induces a non-degenerate pairing on the quotient $W = \widetilde{W}/\widetilde{W}_0$.

Let π be an everywhere unramified cuspidal automorphic representation of G with coefficient field E_π, and let $\lambda_\pi: \mathcal{H} \to E_\pi$ be the associated character. The kernel of λ_π defines a maximal ideal $\mathfrak{m}_\pi$ of $\mathcal{H}$. In Theorem 1.1, [61], the authors show W admits the spectral decomposition

$$W = W_{\mathrm{Eis}} \oplus \left(\bigoplus_\pi W_\pi \right)$$

where π runs over the set of everywhere unramified cuspidal automorphic representations of G and the pieces in the decomposition are defined as the eigenspaces

$$W_\pi = W[\mathfrak{m}_\pi] = \{ w \in W : \mathfrak{m}_\pi \cdot w = 0 \}$$
$$W_{\mathrm{Eis}} = W[\mathcal{I}_{\mathrm{Eis}}] = \{ w \in W : \mathcal{I}_{\mathrm{Eis}} \cdot w = 0 \}.$$

In this version of the decomposition, the theorem we obtain is the following.

Theorem 4.47 (Theorem 1.2, [61]). *Let π be an everywhere unramified cuspidal automorphic representation of G and let $[\mathrm{Sht}_T^r]_\pi \in W_\pi$ be the π-isotypic piece of $\theta_*^r[\mathrm{Sht}_T^r] \in W$. Then we have an equality*

$$\frac{1}{2(\log q)^r} |\omega_X| \mathcal{L}^{(r)}(\pi_{F'}, 1/2) = \langle [\mathrm{Sht}_T^r]_\pi, [\mathrm{Sht}_T^r]_\pi \rangle_{\mathrm{Sht}_G^{\prime r}, \pi},$$

where ω_X is the canonical divisor of X and $|\omega_X| = q^{-\deg(\omega_X)}$.

The second approach to the spectral decomposition is to consider the cycle class of the Heegner-Drinfeld cycle in the ℓ-adic cohomology for a prime ℓ different from p. For this we consider the middle degree cohomology with compact support

$$V'_{\mathbb{Q}_\ell} = H^{2r}_c((\mathrm{Sht}'^r_G) \otimes_k \overline{k}, \mathbb{Q}_\ell)(r).$$

As a $\mathbb{Q}_\ell$-vector space it is also endowed with a cup product $(\cdot, \cdot) : V'_{\mathbb{Q}_\ell} \times V'_{\mathbb{Q}_\ell} \to \mathbb{Q}_\ell$ which serves as the analogue of the intersection pairing. In Theorem 1.5, [61], the authors obtain the spectral decomposition of $\mathcal{H}_{\mathbb{Q}_\ell}$-modules

$$V'_{\mathbb{Q}_\ell} = V'_{\mathbb{Q}_\ell, \mathrm{Eis}} \oplus \left(\bigoplus_{\mathfrak{m}} V'_{\mathbb{Q}_\ell, \mathfrak{m}} \right)$$

where the sum is over a finite set of maximal ideals $\mathfrak{m} \subseteq \mathcal{H}_{\mathbb{Q}_\ell}$ whose residue fields $E_{\mathfrak{m}} = \mathcal{H}_{\mathbb{Q}_\ell}/\mathfrak{m}$ are finite extensions of $\mathbb{Q}_\ell$, and the pieces in the decomposition are the generalized eigenspaces

$$V'_{\mathbb{Q}_\ell, \mathfrak{m}} = \bigcup_{i>0} V'_{\mathbb{Q}_\ell}[\mathfrak{m}^i] = \bigcup_{i>0} \{ v \in V'_{\mathbb{Q}_\ell} : \mathfrak{m}^i \cdot v = 0 \}$$

$$V'_{\mathbb{Q}_\ell, \mathrm{Eis}} = \bigcup_{i>0} V'_{\mathbb{Q}_\ell}[\mathcal{I}^i_{\mathrm{Eis}}] = \bigcup_{i>0} \{ v \in V'_{\mathbb{Q}_\ell} : \mathcal{I}^i_{\mathrm{Eis}} \cdot v = 0 \}.$$

This decomposition is analogous to the previous cycle-theoretic version as there the generalized eigenspaces of W coincide with the eigenspaces as W is a cyclic module over $\mathcal{H}$. However, unlike the previous cycle-theoretic version, we cannot be sure all terms in this decomposition are automorphic.

Given an everywhere unramified cuspidal automorphic representation π of G with coefficient field E_π, we identify maximal ideals $\mathfrak{m}_{\pi, \lambda}$ with automorphic origins appearing in the decomposition in the following way. Given a character $\lambda_\pi : \mathcal{H} \to E_\pi$, we may extend the coefficients to $\mathbb{Q}_\ell$ to get the ℓ-adic character

$$\lambda_\pi \otimes \mathbb{Q}_\ell : \mathcal{H}_{\mathbb{Q}_\ell} \to E_\pi \otimes \mathbb{Q}_\ell \simeq \prod_{\lambda | \ell} E_{\pi, \lambda}.$$

The λ in the decomposition run over all places of E_π over ℓ. Let $\mathfrak{m}_{\pi, \lambda}$ be the kernel of the λ-component $\mathcal{H}_{\mathbb{Q}_\ell} \to E_{\pi, \lambda}$ of the above decomposition; this defines a maximal ideal in $\mathcal{H}_{\mathbb{Q}_\ell}$. Then $V'_{\pi, \lambda} := V'_{\mathbb{Q}_\ell, \mathfrak{m}_{\pi, \ell}}$ occurs in the decomposition and is automorphic. We fix one such place λ and denote the $E_{\pi, \lambda}$-linear pairing on $V'_{\pi, \lambda}$ coming from the restriction of the cup product pairing by

$$(\cdot, \cdot)_{\pi, \lambda} : V'_{\pi, \lambda} \times V'_{\pi, \lambda} \to E_{\pi, \lambda}.$$

The theorem we obtain in this setting is the following.

Theorem 4.48 (Theorem 1.6, [61]). *Let π be an everywhere unramified cuspidal automorphic representation of G. Let $[\mathrm{Sht}_T^r]_{\pi,\lambda} \in V'_{\pi,\lambda}$ be the projection of the cycle class $\mathrm{cl}(\theta_*^r[\mathrm{Sht}_T^r]) \in V'_{\overline{\mathbb{Q}}_\ell}$ to the summand $V_{\pi,\lambda}$. Then we have an equality*

$$\frac{1}{2(\log q)^r}|\omega_X|\mathcal{L}^{(r)}(\pi_{F'},1/2) = ([\mathrm{Sht}_T^r]_{\pi,\lambda}, [\mathrm{Sht}_T^r]_{\pi,\lambda})_{\pi,\lambda},$$

where ω_X is the canonical divisor of X and $|\omega_X| = q^{-\deg(\omega_X)}$.

The key identity.   In view of the spectral decompositions of both $\mathbb{I}_r(f)$ and $\mathbb{J}_r(f)$, to prove Theorem 4.32 (or each of the precise versions Theorem 4.47 and Theorem 4.48), it remains to compare the terms in their orbital expansions indexed by $u \in \mathbb{P}^1(F) - \{1\}$:

$$\mathbb{I}_r(u,f) = (\log q)^{-r}\mathbb{J}_r(u,f), \quad f \in \mathcal{H}.$$

Establishing this key identity is a very involved algebro-geometric argument in [61], so we omit most details in commenting on its proof.

The first part of the argument is to reinterpret the orbital integral expansion of $\mathbb{J}_r(f)$ as a certain weighted count of effective divisors on the curve X. For this we introduce the global moduli space of orbital integrals and construct over it a local system whose Lefschetz trace computes the orbital decomposition of $\mathbb{J}_r(f)$.

First we define the base $\mathcal{A}_d$ to be the moduli stack of triples (Δ, a, b) where $\Delta \in \mathrm{Pic}_X^d$ and a and b are sections of Δ with the open condition that they are not simultaneously zero. For an effective divisor $D \subseteq X$ of degree $d \geq 0$, we let $\mathcal{A}_D \subseteq \mathcal{A}_d$ be the substack that classifies triples $(\mathcal{O}_X(D), a, b)$ with the condition that $a - b$ is the tautological section $1 \in \Gamma(X, \mathcal{O}_X(D))$. We remark that such a triple is determined uniquely by the section $a \in \Gamma(X, \mathcal{O}_X(D))$, and so we get a canonical isomorphism

$$\mathcal{A}_D \simeq \Gamma(X, \mathcal{O}_X(D)).$$

On the level of k-points, this induces the map

$$\mathrm{inv}_D : \mathcal{A}_D(k) \simeq \Gamma(X, \mathcal{O}_X(D)) \hookrightarrow \mathbb{P}^1(F) - \{1\}$$

$$(\mathcal{O}_X(D), a, b) \leftrightarrow a \mapsto (a-1)/a - 1 - a^{-1}.$$

Let $\underline{d}$ be a quadruple of non-negative integers $(d_{ij})_{i,j \in \{1,2\}}$ satisfying $d_{11} + d_{12} = d_{12} + d_{21} = d$. The moduli stack of orbital integrals $\mathcal{N}_{\underline{d}}$ classifies quintuples $(\mathcal{K}_1, \mathcal{K}_2, \mathcal{K}'_1, \mathcal{K}'_2, \phi)$ up to Pic_X-equivalence where

- $\mathcal{K}_i, \mathcal{K}'_i \in \mathrm{Pic}_X$ with $\deg \mathcal{K}'_i - \deg \mathcal{K}_j = d_{ij}$,
- $\phi : \mathcal{K}_1 \oplus \mathcal{K}_2 \to \mathcal{K}'_1 \oplus \mathcal{K}'_2$ is an $\mathcal{O}_X$-linear map:

$$\phi = \begin{pmatrix} \phi_{11} & \phi_{12} \\ \phi_{21} & \phi_{22} \end{pmatrix}$$

where $\phi_{ij} : \mathcal{K}_j \to \mathcal{K}'_i$, and

- if $d_{11} < d_{22}$ then $\phi_{11} \neq 0$ and if $d_{12} < d_{21}$ then $\phi_{12} \neq 0$; moreover, at most one of the four maps ϕ_{ij} can be zero.

There is a map

$$f_{\mathcal{N}_{\underline{d}}} : \mathcal{N}_{\underline{d}} \to \mathcal{A}_d$$

$$(\mathcal{K}_1, \mathcal{K}_2, \mathcal{K}'_1, \mathcal{K}'_2, \phi) \mapsto (\mathcal{K}'_1 \otimes \mathcal{K}'_2 \otimes \mathcal{K}_1^{-1} \otimes \mathcal{K}_2^{-1}; \phi_{11} \otimes \phi_{22}; \phi_{12} \otimes \phi_{21}).$$

In §3.3.1 [61], the authors construct the local system $\mathbf{R}f_{\mathcal{N}_{\underline{d}},*}L_{\underline{d}}$ on $\mathcal{A}_d$ for which they show:

Theorem 4.49 (Corollary 3.3, [61]). *For $D \subseteq X(k)$ an effective divisor of degree $d \geq 0$ and $u \in \mathbb{P}^1(F) - \{1\}$, we have*

$$\mathbb{J}_r(u, h_D) = \begin{cases} (\log q)^r \sum_{\underline{d} \in \Sigma_d} (2d_{12} - d)^r \\ \qquad \mathrm{Tr}\left(\mathrm{Frob}_a; (\mathbf{R}f_{\mathcal{N}_d,*}L_{\underline{d}})_{\overline{a}}\right) & \text{if } u = \mathrm{inv}_D(a),\, a \in \mathcal{A}_D(k), \\ 0 & \text{otherwise.} \end{cases}$$

Here, Σ_d is the set of quadruples of non-negative integers $\underline{d} = (d_{ij})_{i,j \in \{1,2\}}$ satisfying $d_{11} \mid d_{12} = d_{12} + d_{21} - d$.

The intersection number $\mathbb{I}_r(f)$ can also be expressed as the trace of a correspondence acting on the cohomology of a certain stack which can be viewed as a moduli space of naive shtuka. This is the stack $\mathcal{M}_d$ classifying quadruples $(\mathcal{L}, \mathcal{L}', \alpha, \beta)$ up to Pic_X-equivalence where

- $\mathcal{L}, \mathcal{L}' \subset \mathrm{Pic}(X' \times S)$ such that $\deg(\mathcal{L}'_s) - \deg(\mathcal{L}_s) = d$ for all geometric points $s \in S$,
- $\alpha : \mathcal{L} \to \mathcal{L}'$ is an $\mathcal{O}_{X'}$-linear map,
- $\beta : \mathcal{L} \to \sigma^* \mathcal{L}'$ is an $\mathcal{O}_{X'}$-linear map, and
- for each geometric point $s \in S$, the restrictions $\alpha|_{X' \times s}$ and $\beta|_{X' \times s}$ are not both zero.

There is a morphism

$$f_{\mathcal{M}} : \mathcal{M}_d \to \mathcal{A}_d$$

$$(\mathcal{L}, \mathcal{L}', \alpha, \beta) \mapsto (\mathrm{Nm}(\mathcal{L}') \otimes \mathrm{Nm}(\mathcal{L}^{-1}), \mathrm{Nm}(\alpha), \mathrm{Nm}(\beta)).$$

We let $\mathcal{A}_d^\diamond \subseteq \mathcal{A}_d$ be the open subset consisting of (Δ, a, b) where $b \neq 0$, and $\mathcal{M}_d^\diamond$ be the preimage under $f_{\mathcal{M}}$ of this open locus.

In §6.2.1 [61] the authors define the general notion of self-correspondence of $\mathcal{M}_d$ over $\mathcal{A}_d$. For application, we need only consider the correspondence $\mathcal{H}^\circ$ on the open locus $\mathcal{M}_d^\circ$, which is a stack classifying pairs of commutative diagrams

$$
\begin{array}{ccc}
\mathcal{L}_0 & \xrightarrow{f} & \mathcal{L}_1 \\
\downarrow{\scriptstyle\alpha_0} & & \downarrow{\scriptstyle\alpha_1} \\
\mathcal{L}_0' & \xrightarrow{f'} & \mathcal{L}_1'
\end{array}
\qquad\qquad
\begin{array}{ccc}
\mathcal{L}_0 & \xrightarrow{f} & \mathcal{L}_1 \\
\downarrow{\scriptstyle\beta_0} & & \downarrow{\scriptstyle\beta_1} \\
\sigma^*\mathcal{L}_0' & \xrightarrow{\sigma^*f'} & \sigma^*\mathcal{L}_1'
\end{array}
$$

where $(\mathcal{L}_i, \mathcal{L}_i', \alpha_i, \beta_i) \in \mathcal{M}_d^\circ$ for $i = 1, 2$ and the cokernel of f and f' are invertible sheaves supported at the same point $x' \in X'$. Using the Lefschetz trace formula, the authors arrive at the following:

Theorem 4.50 (Theorem 6.5, [61]).   *Let $D \subseteq X$ be an effective divisor of degree $d \geq \max\{2g' - 1, 2g\}$. Then we have*

$$
\mathbb{I}_r(h_D) = \sum_{a \in \mathcal{A}_D(k)} \mathrm{Tr}\left((f_{\mathcal{M},!}[\mathcal{H}^\circ])_a^r \circ \mathrm{Frob}_a; (\mathbf{R}f_{\mathcal{M},!}\mathbb{Q}_\ell)_{\bar{a}}\right).
$$

In particular, this gives the decomposition

$$
\mathbb{I}_r(h_D) = \sum_{u \in \mathbb{P}^1(F) - \{1\}} \mathbb{I}_r(u, h_D)
$$

where

$$
\mathbb{I}_r(u, h_D) = \begin{cases} \mathrm{Tr}\left((f_{\mathcal{M},!}[\mathcal{H}^\circ])_a^r \circ \right. \\ \quad \left. \mathrm{Frob}_a, (\mathbf{R}f_{\mathcal{M},!}\mathbb{Q}_\ell)_{\bar{a}}\right) & \text{if } u = \mathrm{inv}_D(a) \text{ for some } a \in \mathcal{A}_D(k), \\ 0 & \text{otherwise.} \end{cases}
$$

To prove the key identity now essentially reduces to show the local terms in the two trace formulae of Theorem 4.49 and Theorem 4.50 are equal. To make this comparison makes heavy use of the *perverse continuation principle*, by which if we view $\mathbf{R}f_{\mathcal{M},!}\mathbb{Q}_\ell$ and $\mathbf{R}f_{\mathcal{N}_d,*}L_{\underline{d}}$ as (shifted) perverse sheaves satisfying certain properties, then establishing an identity between them over a suitable open locus shows an identity globally. This principle carries technical significance, as the geometry of the relevant moduli spaces is well-understood over an open locus but not everywhere. Proposition 8.2 and Proposition 8.5 in [61] give the isomorphisms between each of $\mathbf{R}f_{\mathcal{M},!}\mathbb{Q}_\ell$ and $\mathbf{R}f_{\mathcal{N}_d,*}L_{\underline{d}}$ and their perverse structures $\bigoplus_{i,j=0}^{d}(K_i \boxtimes K_j)|_{\mathcal{A}_d}$ and $(K_{d_{11}} \boxtimes K_{d_{12}})|_{\mathcal{A}_d}$, respectively. Proposition 8.3 computes the weight factors of the action of the correspondence $f_{\mathcal{M},!}[\mathcal{H}^\circ]$ on each piece of the perverse sheaf $\bigoplus_{i,j=0}^{d}(K_i \boxtimes K_j)|_{\mathcal{A}_d}$, which allows one to compare the two trace formulae and establish the key identity for most elements in $\mathcal{H}$ in Theorem 8.1. The proof of the key identity for all $f \in \mathcal{H}$ is concluded in Theorem 9.2.

BIBLIOGRAPHY

[1] G. Anderson: *Logarithmic derivatives of Dirichlet L-functions and the periods of abelian varieties*, Compos. Math. **45** (1982), 315-332.

[2] Y. André: *Finitude des couples d'invariants modulaires singuliers sur une courbe algébrique plane non modulaire*, J. Reine Angew. Math. **505** (1998), 203–208.

[3] F. Andreatta, E. Goren, B. Howard, and K. Madapusi Pera: *Faltings heights of abelian varieties with complex multiplication*, Annals of Math. **187** (2018), 391–531.

[4] S. J. Arakelov: *Families of algebraic curves with fixed degeneracies*, Izv. Akad. Nauk SSSR Ser. Mat. **35** (1971), no. 6, 1269-1293.

[5] Y. Bilu, F. Luca, and D. Masser: *Collinear CM-points*, Algebra Number Theory, vol. 11, no. 5 (2017), 1047–1087.

[6] Y. Bilu, D. Masser, and U. Zannier: *An effective Theorem of André for CM-points on a plane curve*, Math. Proc. Cambridge Philos. Soc. **154** (2013), 145–152.

[7] H. Carayol: *Sur la mauvais réduction des courbes de Shimura*, Compos. Math. **59** (1986), 151-230.

[8] C-L. Chai, B. Conrad, and F. Oort: *Complex Multiplication and Lifting Problems*, American Mathematical Society (2014).

[9] S. Chowla: *On a conjecture of Marshall Hall*, Proc. Nat. Acad. Sci. **56** (1966), 417-418.

[10] S. Chowla: *Remarks on class-invariants and related topics, chapter VI of Seminar on Complex Multiplication* (eds. A. Borel et al.) Lect. Notes Math. **21** (1966).

[11] S. Chowla and A. Selberg: *On Epstein's zeta function. I*, Proceedings of the National Academy of Sciences of the United States of America **35** (1949), 371-374.

[12] S. Chowla and A. Selberg: *On Epstein's Zeta-function*, Journal für die reine und angewandte Mathematik **227** (1967), 86-110.

[13] P. Colmez: *Périodes des variétés abéliennes à multiplicaton complexe*, Annals of Math. **138** (1993), 625-683.

[14] P. Deligne: *Preuve des conjectures de Tate et de Shafarevitch (d'aprés G. Faltings)*, Astérisque **121-122** (1985), 25-42.

[15] P. Deligne: *Un théorème de finitude pour la monodromie*, In Discrete groups in geometry and analysis (New Haven, Conn., 1984), vol. 67 of Progr. Math., Birkhäuser, Boston (1987), 1-19.

[16] F. Diamond and J. Shurman: *A first course in modular forms*, Graduate Texts in Mathematics **228**, Springer-Verlag, New York (2005).

[17] V. G. Drinfeld: *Moduli varieties of F-sheaves*, Funktsional. Anal. i Prilozhen. **2** (1987), 23-41.

[18] B. Edixhoven, B. Moonen, and F. Oort: *Open problems in algebraic geometry*, Bull. Sci. Math. **125** (2001), 122.

[19] G. Faltings: *Endlichkeitssätze für abelsche Varietäten über Zahlkörpern*, Invent. Math. **73** (1983), no. 3, 349-366.

[20] G. Faltings: *Arakelov's theorem for abelian varieties*, Invent. Math. **73** (1983), no. 3, 337-347.

[21] G. Faltings: *Calculus on arithmetic surfaces*, Ann. of Math. **119** (1984), 387-424.

[22] G. Faltings and C.-L. Chai: *Degeneration of abelian varieties*, Ergebnisse der Mathematik und ihrer Grenzgebiete, 3. Folge, vol. 22, Springer-Verlag (1990).

[23] G. Frey: *Links between solutions of $A - B = C$ and elliptic curves*, In Number theory (Ulm, 1987), vol. 1380 of Lecture Notes in Math., Springer, New York (1989), 31–62.

[24] Z. Gao: *Towards the André-Oort conjecture for mixed Shimura varieties: the Ax-Lindemann theorem and lower bounds for Galois orbits of special points*, J. Reine Angew. Math. **732** (2017), 85-146.

[25] Z. Gao: *About the mixed André-Oort conjecture: reduction to a lower bound for the pure case*, C. R. Math. Acad. Sci. Paris **354** (2016), 659-663.

[26] A. Granville and H. Stark: *ABC implies no "Siegel zeros" for L-functions of characters with negative discriminant*, Invent. Math. **139** (2000), 509–523.

[27] B. H. Gross: *On the periods of abelian integrals and a formula of Chowla and Selberg,* Invent. Math. **45** (1978), 193–211.

[28] B. H. Gross and D. Zagier: *Heegner points and derivatives of L-series*, Invent Math. **84** (1986), 225–320.

[29] R. von Känel: *On Szpiro's discriminant conjecture*, Int. Math. Res. Not. **16** (2014), 4457–4491.

[30] B. Klingler, E. Ullmo, and A. Yafaev: *The hyperbolic Ax-Lindemann-Weierstrass conjecture*, Inst. Hautes Études Sci. Publ. Math. **123** (2016), 333-360.

[31] A. Kresch: *Cycle groups for Arin stacks,* Invent. Math. **138** (1991), no. 3, 495–536.

[32] L. Kühne: *An effective result of André–Oort type*, Ann. Math. (2) **176** (2012), 651-671.

[33] L. Kühne: *An effective result of André–Oort type II*, Acta Arith. **161** (2013), 1–19.

[34] Q. Liu, D. Lorenzini, and M. Raynaud: *Néron models, Lie algebras, and reduction of curves of genus one*, Invent. Math. **157** (2004), no. 3, 455–518.

[35] K. Mahler: *On Heckes Theorem on the real zeros of the L-functions and the class number of quadratic fields*, J. London Math. Soc. **9** (1934), 298-302.

[36] M. Lerch: *Sur quelques formules relatives au nombre des classes*, Bulletin des Sciences Mathématiques **21** (1897), 290–304.

[37] D. W. Masser: *Note on a conjecture of Szpiro*, Astérisque **183** (1990), 19-23.

[38] D. W. Masser: *On abc and discriminants*, Proc. Amer. Math. Soc. **130** (2002), no. 11, 3141–3150.

[39] D. W. Masser and G. Wüstholz: *Factorization estimates for abelian varieties*, Inst. Hautes Études Sci. Publ. Math. **81** (1995), 5–24.

[40] L. Mocz: *A New Northcott Property for Faltings Height*, preprint, arXiv:1709.06098 (2017).

[41] L. Moret-Bailly: *La formule de Noether pour les surfaces arithmétiques*, Inv. Math. **98** (1989), 491-498.

[42] A. Obus: *On Colmez's product formula for periods of CM-abelian varieties*, Math. Ann. **356** (2013), 401-418.

[43] A. N. Paršin: *Algebraic curves over function fields. I*, Izv. Akad. Nauk SSSR Ser. Mat. **32** (1968), 1191-1219.

[44] J. Pila: *O-minimality and the André-Oort conjecture for $\mathbb{C}^N$*, Annals of Math. **173** (2011), 1779-1840.

[45] J. Pila and J. Tsimerman: *Ax–Lindemann for $\mathcal{A}_g$*, Annals of Math. **179** (2014), 659–681.

[46] J. Pila and J. Wilkie: *The rational points of a definable set*, Duke Math. J. **133** (2006), 591–616.

[47] J. Pila and U. Zannier: *Rational points in periodic analytic sets and the Manin-Mumford conjecture*, Rend. Mat. Acc. Lincei **19** (2008), 149–162.

[48] L. Szpiro: *Sur la théorème de rigidité de Parshin et Arakelov*, Astérisque **64** (1979), 169-202.

[49] L. Szpiro: *Discriminant et conducteur des courbes elliptiques*, Astérisque **183** (1990), 7–18.

[50] J. Tsimerman: *A proof of the André–Oort conjecture for A_g*, Annals of Math. **187** (2018), 379-390.

[51] E. Ullmo: *Applications du théorème d'Ax-Lindemann hyperbolique*, Compos. Math. **150** (2014), 175-190.

[52] E. Ullmo and A. Yafaev: *Hyperbolic Ax-Lindemann theorem in the cocompact case*, Duke Math. J. **163** (2014), 433-463.

[53] Y. Varshavsky: *Moduli spaces of principal F-bundles*, Selecta Math. (N.S.)
      10 (2004), 131–166.

[54] A. Weil: *Elliptic functions according to Eisenstein and Kronecker*, Ergeb-
      nisse der Math. und ihrer Grenzgebiete **88**, Springer-Verlag (1976).

[55] G. Wüstholz: *A note on the conjectures of André–Oort and Pink with an
      appendix by Lars Kühne*, Bull. Inst. Math. Acad. Sin. (N.S.) (9) **4** (2014),
      735–779.

[56] T. Yang: *Chowla–Selberg formula and Colmez's conjecture*, Canad. J. Math.
      62 (2) (2010), 456–472.

[57] T. Yang: *An arithmetic intersection formula on Hilbert modular surfaces*,
      Amer. J. Math. **132** (2010), 1275-1309.

[58] T. Yang: *Arithmetic intersection on a Hilbert modular surface and the Falt-
      ings height*, Asian J. Math. **17** (2013), 335-381.

[59] X. Yuan, S. Zhang, and W. Zhang: *The Gross-Zagier formula on Shimura
      curves*, Annals of Mathematics Studies **184**, Princeton University Press,
      Princeton, NJ, 2013.

[60] X. Yuan and S. Zhang: *On the averaged Colmez conjecture*, Annals of Math.
      187 (2018), 533-638.

[61] Z. Yun and W. Zhang: *Shtukas and the Taylor expansion of L-functions*,
      Annals of Math. **186** (2017), 767-911.

[62] Y. Zarhin: *A finiteness theorem for unpolarized Abelian varieties over num-
      ber fields with prescribed places of bad reduction*, Invent. Math. **79** (1985),
      309–321.

[63] S. Zhang: *Gross–Schoen cycles and dualising sheaves*, Invent Math. **179**
      (2010), 1-73.

[64] S. Zucker: *Hodge Theory with degenerating coefficients: L2 cohomology in
      the Poincare metric*, Annals of Math. **109** (1979), 415–476.

List of Contributors

Laura Capuano University of Oxford, Mathematical Institute, Andrew Wiles Building, Radcliffe Observatory Quarter, Woodstock Road, Oxford, EX2 6GG, UK
laura.capuano@maths.ox.ac.uk

Ziyang Gao CNRS, IMJ-PRG, 4 place de Jussieu, 75005 Paris, France
ziyang.gao@imj-prg.fr

Sergey Gorchinskiy NRU HSI, Laboratory of Mirror Symmetry, 6 Usacheva Str., Moscow, Russia, 119048
gorchins@mi.ras.ru

Peter Jossen ETH Zurich, Department of Mathematics, Raemistrasse 101, 8092 Zurich, Switzerland
peter.jossen@math.ethz.ch

Rafael von Känel Princeton University, Department of Mathematics, Fine Hall, Princeton, NJ 08544-1000, USA
rafaelvonkanel@gmail.com

Christina Karolus University of Salzburg, Department of Mathematics, Hellbrunnerstrasse 34, 5020 Salzburg, Austria
christina.karolus@sbg.ac.at

Lars Kühne University of Basel, Department of Mathematics and Computer Science, Spiegelgasse 1, 4051 Basel, Switzerland
lars.kuehne@unibas.ch

Lucia Mocz Princeton University, Department of Mathematics, Fine Hall, Princeton, NJ 08544-1000, USA
lmocz@math.princeton.edu

Francesco Veneziano Scuola Normale Superiore di Pisa, Piazza dei Cavalieri, 7, 56126 Pisa, Italy
francesco.veneziano@sns.it